Grassroots Environmentalism

Grassroots activism is essential to the success of the contemporary environmental movement, which depends on the organization of local activists as well as state, national, and international organizations. Yet grassroots activists confront numerous challenges as they attempt to organize diverse participants and devise fresh strategies and tactics. Drawing on more than seven years of fieldwork following five organizations in Pittsburgh over time, this book sheds light on the struggles that activists face and the factors that sustain movements. It examines individual motivations and participation, organizational structures and cultures, relationships in movement communities, and strategies and tactics, including issue framing. The book shows that collective action campaigns and tactics generate solidarity, maintain involvement, and bring in new participants even as organizers struggle to devise effective new types of actions.

SUZANNE STAGGENBORG is Professor of Sociology at the University of Pittsburgh. Her research examines the dynamics of social movements. She is the author of *The Pro-Choice Movement: Organization and Activism in the Abortion Conflict* (1991) and *Social Movements* (2016) and is the 2019 winner of the John D. McCarthy Award for Lifetime Achievement in the Scholarship of Social Movements and Collective Behavior.

Cambridge Studies in Contentious Politics

Rina Agarwala, *Informal Labor, Formal Politics, and Dignified Discontent in India*
Ronald Aminzade, *Race, Nation, and Citizenship in Post-Colonial Africa: The Case of Tanzania*
Ronald Aminzade et al., *Silence and Voice in the Study of Contentious Politics*
Javier Auyero, *Routine Politics and Violence in Argentina: The Gray Zone of State Power*
Phillip M. Ayoub, *When States Come Out: Europe's Sexual Minorities and the Politics of Visibility*
Amrita Basu, *Violent Conjunctures in Democratic India*
W. Lance Bennett and Alexandra Segerberg, *The Logic of Connective Action: Digital Media and the Personalization of Contentious Politics*
Nancy Bermeo and Deborah J. Yashar, editors, *Parties, Movements, and Democracy in the Developing World*
Clifford Bob, *The Global Right Wing and the Clash of World Politics*
Clifford Bob, *The Marketing of Rebellion: Insurgents, Media, and International Activism*
Robert Braun *Protectors of Pluralism: Religious Minorities and the Rescue of Jews in the Low Countries during the Holocaust*
Charles Brockett, *Political Movements and Violence in Central America*
Marisa von Bülow, *Building Transnational Networks: Civil Society and the Politics of Trade in the Americas*
Valerie Bunce and Sharon Wolchik, *Defeating Authoritarian Leaders in Postcommunist Countries*
Teri L. Caraway and Michele Ford, *Labor and Politics in Indonesia*

(*continued after Index*)

Grassroots Environmentalism

SUZANNE STAGGENBORG
University of Pittsburgh

CAMBRIDGE
UNIVERSITY PRESS

CAMBRIDGE
UNIVERSITY PRESS

University Printing House, Cambridge CB2 8BS, United Kingdom

One Liberty Plaza, 20th Floor, New York, NY 10006, USA

477 Williamstown Road, Port Melbourne, VIC 3207, Australia

314–321, 3rd Floor, Plot 3, Splendor Forum, Jasola District Centre, New Delhi – 110025, India

79 Anson Road, #06–04/06, Singapore 079906

Cambridge University Press is part of the University of Cambridge.

It furthers the University's mission by disseminating knowledge in the pursuit of education, learning, and research at the highest international levels of excellence.

www.cambridge.org
Information on this title: www.cambridge.org/9781108478489
DOI: 10.1017/9781108777872

First published 2020

A catalogue record for this publication is available from the British Library.

Library of Congress Cataloging-in-Publication Data
NAMES: Staggenborg, Suzanne, author.
TITLE: Grassroots environmentalism / Suzanne Staggenborg.
DESCRIPTION: Cambridge, United Kingdom ; New York, NY : Cambridge University Press, 2020. | SERIES: Cambridge studies in contentious politics | Includes bibliographical references and index.
IDENTIFIERS: LCCN 2020021919 (print) | LCCN 2020021920 (ebook) | ISBN 9781108478489 (hardback) | ISBN 9781108745864 (paperback) | ISBN 9781108777872 (epub)
SUBJECTS: LCSH: Environmentalism.
CLASSIFICATION: LCC GE195 .S73 2020 (print) | LCC GE195 (ebook) | DDC 363.7–dc23
LC record available at https://lccn.loc.gov/2020021919
LC ebook record available at https://lccn.loc.gov/2020021920

ISBN 978-1-108-47848-9 Hardback
ISBN 978-1-108-74586-4 Paperback

To Charlie and Laura

Contents

Figures

Tables

Acknowledgments

This book brings together my longstanding interests in environmental problems, movement dynamics, and fieldwork methods. The climate crisis motivated me to devote my research efforts to examining how environmental activism can help save our planet. My interest in organizational dynamics and strategic choices led me to focus on the internal workings of the movement. To be sure, this research is only a very partial contribution to our understanding of the movement's potential. National and transnational movement organizations and actions are certainly essential, but so are local groups; I am particularly interested in what ordinary people can accomplish by participating in grassroots organizations and actions. By studying the environmental movement in Pittsburgh through participant observation, I was able to see firsthand both how much they can accomplish and how many difficulties they face in trying to organize and create change through collective action. I was happy to be able to make small contributions to the activities of the groups I studied, something I would not have had time to do if it was not also a research project.

Grassroots Environmentalism has been many years in the making, and numerous people helped me along the way. My ability to observe movement dynamics depended on my access to local groups. I owe a great debt to the activists who trusted me to be part of their groups, knowing I was studying the movement and not only participating in it. I am also thankful to those who shared their experiences and assessments with me in interviews and in numerous informal conversations. I benefited greatly from the interest, insights, and encouragement that I received from them. Some of these people are named, with their permission, in the book, but there are many unnamed people who also talked to me and helped me understand the movement. I am also grateful to the photographers, credited under their photos, for sharing their pictures with me. Special thanks to Tom Jefferson, a professional photographer who often covers movement events, for graciously allowing me to use his photos.

I appreciate all of the support provided to me by so many activists, and I value the ongoing relationships I have with many of them.

I also received a lot of support from academic sources. The best perk of being chair of my department for six years was that a research assistant came with the job. Hatem Hassan, Matthew Hemphill, Corinne Ogrodnik, and Caitlin Schroering all contributed to this project in various ways. Hatem was my first assistant, early in the project, who assisted with bibliographic work and helped me think through the importance of movement communities in an essay we wrote together. Matt helped me with various tasks, most importantly coding some of the data when I finally admitted to myself that I couldn't do it all alone. He encouraged me greatly when he told me how interesting my fieldnotes were at a time when I was beginning to fear that I was drowning in so much detail. Corinne was my assistant for two years, and she contributed significantly to the project by coding data, interviewing most of the informants from Transition, and constructing the figures on activities in the anti-fracking movement over time with data compiled from email communications. In writing an article together on Transition, we worked through many ideas about the struggling organization. Caitlin was also my assistant for two years, mostly during the time that I was writing the book, and she helped me in too many ways to list. Most importantly, she read and commented on multiple versions of every chapter of the manuscript, providing me with invaluable comments and discussing the work with me as a trusted assessor. I also benefited from hearing about her research on conflicts over water in Pittsburgh and Brazil and from our work together on a paper comparing coalitions that we studied. I am grateful to the Dietrich School of Arts and Sciences at the University of Pittsburgh for funding these research assistants.

In addition to my research assistants, some wonderful colleagues helped me to develop the work. At the University of Pittsburgh, Josh Bloom read an early draft of the entire manuscript and gave me very helpful comments on it. Dana Moss read parts of the manuscript and helped me think it through. I was also fortunate to have excellent reviewers of the manuscript. Paul Lichterman, Nancy Whittier, Shannon Bell, and an anonymous reviewer all gave me extensive comments that aided me greatly as I worked to reorganize and develop the manuscript. I also benefited from comments made by other scholars at professional meetings where I presented early versions of the work. Tony Ladd gave me very useful comments on a chapter I published in his edited book, *Fractured Communities*, which contains material that is reworked across several chapters in this book. Finally, Rod Nelson read as many drafts of every chapter as Caitlin Schroering did, often going over it with me line by line. As a sociologist outside of the social movement field, Rod pushed me to clarify discussions that were confusing to nonspecialists and to eliminate unnecessary jargon. The book is much clearer as a result.

At Cambridge University Press, my first editor, Robert Dreesen, gave me sound advice and encouragement early on, and my next editor, Sara Doskow, along with editorial assistant Cameron Daddis, guided me through the final stages. David Meyer also supported the project as lead editor of the Contentious Politics series. I thank everyone profusely.

Abbreviations and Acronyms

ACHD	Allegheny County Health Department
AG	Allegheny Group (Sierra Club)
CC	Coordinating Committee (Marcellus Protest)
CCJ	Center for Coalfield Justice
CELDF	Community Environmental Legal Defense Fund
DEP	Department of Environmental Protection (Pennsylvania)
ESAT	Endangered Species Action Team (Allegheny Group of Sierra Club)
ExCom	Executive Committee (Sierra Club)
GA	General Assembly
GASP	Group Against Smog and Pollution
GWAT	Global Warming Action Team (Allegheny Group of Sierra Club)
LPAG	Lincoln Place Action Group
LULU	locally unwanted land use
MOB	Marcellus Outreach Butler
MP	Marcellus Protest
NAACP	National Association for the Advancement of Colored People
NIABY	not in anyone's back yard
NIMBY	not in my back yard
NOW	National Organization for Women
PGRP	Pittsburgh G20 Resistance Project
POP	Protect Our Parks
PSEC	Pittsburgh Student Environmental Coalition
RAN	Rainforest Action Network
SC	Sierra Club
SHAADD	South Hills Area Against Dangerous Drilling
SMC	Social Movement Community
SMO	Social Movement Organization
T4T	Training for Transition
TMC	Thomas Merton Center
TRRT	Three Rivers Rising Tide

1

Introduction

In the summer of 2014, representatives of several environmental groups in Pittsburgh began meeting to plan for local participation in the People's Climate March to be held in New York City on September 21, 2014. The Sierra Club, 350.org, and other national and international groups were organizing the march around a United Nations (UN) meeting on climate change held in New York in advance of the Climate Change Conference in Paris in 2015. Across the country, grassroots groups, including affiliates of Rising Tide North America and the Climate Justice Alliance, were organizing to attend the People's Climate March and to participate in a week of activities that would include a nonviolent direct action in New York's financial district to "flood Wall Street" on the day following the main march. Radical and mainstream environmentalists alike were eager to be part of these events, and participants in the Pittsburgh meetings included a staff member of the Sierra Club and members of the Global Warming Action Team of the Allegheny Group of the Sierra Club, local anti-fracking and environmental justice groups, student environmental groups, and a community organization called Action United. In addition to organizing Pittsburgh residents to attend the People's Climate March, their goal was to build a diverse local climate movement consisting not only of environmentalists but also unions, faith leaders, students, and residents of disadvantaged communities. On September 8, 2014, this ad hoc group held a "pre-event" with speakers, music, and refreshments that attracted a diverse crowd of people. A national organizer working with 350.org and Rising Tide came from New York to speak about the march, followed by local speakers from a student coalition, the local Service Employees International Union (SEIU), the American Friends Service Committee (AFSC), and the Sierra Club among others. Framing of the issue reflected the diversity of the event coalition; the SEIU speaker talked about how low-income workers are "extremely affected by climate change," and the AFSC representative talked about "climate justice" in connection with racial justice and human rights (fieldnotes, September 8, 2014). A few weeks later, several hundred Pittsburgh residents traveled to New York in at least five buses organized by local groups.

Local organizing for the People's Climate March exemplifies both the potential and difficulties of grassroots environmentalism. The diverse group of organizers included staff members and volunteers, members of groups with different ideological and strategic approaches, and people of different class and racial backgrounds. They had resources to offer, such as subsidies for low-income participants, and network connections to many potential constituents for a growing climate movement. The organizers did some very effective local outreach to encourage people to travel to New York, but they failed to use this organizing work to jumpstart a local climate movement and build a diverse coalition. That would have required more leadership and organization, as well as ongoing strategies and tactics, than organizers were ready to provide. Some people who traveled to New York for the People's Climate March, and to an earlier demonstration against the Keystone XL pipeline in Washington, DC in February 2013, did talk together on their bus rides and eventually founded a chapter of 350.org in Pittsburgh. It was not easy, however, to build a movement of diverse participants, and it was difficult to keep up the momentum following the brief organizing campaign around the People's Climate March. Nevertheless, it was clear that many local activists cared deeply about climate change, which is strongly linked both to other environmental issues such as shale gas drilling and to concerns about social justice and inequality.

Indeed, global warming is the overriding concern of the environmental movement today, and the climate crisis makes environmentalism arguably the most important social movement of our time. The contemporary environmental movement has had an important impact, both in the United States and around the world, but it has also fallen short in many ways – most notably in failing to focus soon enough on climate change. The movement has been criticized for being overly institutionalized, consisting of large national organizations that are too limited in their strategies and tactics and too beholden to elite interests. Grassroots environmentalism and direct-action tactics are often viewed as the solution to these limitations; a contentious, visible, and persistent movement involving a large number of activists may succeed in creating change where a mainstream one consisting of large organizations with a limited range of strategies has failed. Large national and international organizations might stifle mass organizing, but the combination of grassroots activism and national and international organizations could also be a winning one. In the case of the People's Climate March, national and local groups collaborated on a very successful demonstration despite the lack of follow-up organizing in Pittsburgh. In 2017, national and grassroots activists organized another successful People's Climate March in Washington, DC and in cities across the country to protest the antienvironmental Trump administration.

My goal in this book is to examine both the promise of grassroots environmentalism and the challenges facing this wing of the movement. My account draws on more than seven years of participant observation research in the local environmental movement in Pittsburgh. It examines the interactions of individuals and organizations in the local movement community and details participation in the movement by five organizations, which vary by structure and culture. Two are part of larger structures: Transition Pittsburgh was a local initiative of the international Transition organization, which promotes sustainable communities. The Allegheny Group of the Sierra Club is one of ten regional groups making up the Pennsylvania chapter of the national Sierra Club. The other three groups are local ones that were part of the movement against shale gas drilling in Pittsburgh and Pennsylvania: the Shadbush Environmental Justice Collective, Marcellus Protest, and Protect Our Parks. In following these movement groups and activities over a number of years, I provide a detailed account of different facets of grassroots environmentalism and show why and how local residents sometimes mobilize to protect the environment and what they are able to accomplish.

Grassroots environmentalism involves a variety of movement supporters with different backgrounds and ideologies in democratic collective action. Some are long-time progressive activists who come to the environmental movement after working for other causes such as peace and feminism, dating back to the 1960s. Many experienced environmentalists become involved in campaigns that have emerged more recently, such as the battle against shale gas drilling (known as "fracking"). Other participants are much newer to movement activism in general and environmentalism in particular, becoming involved for the first time in response to threats posed by industrial activities such as strip mining and shale gas and oil drilling. Some are young people aroused by environmental outrages such as mountaintop removal, fracking, and climate change. Others are relatively apolitical green entrepreneurs who simply want to create solutions to climate change and other problems. Many retirees, some with experience in liberal causes, find environmental groups and campaigns when they look for meaningful places to use their skills and resources. The movement organizations that attract grassroots environmentalists are equally varied. They include long-standing, mainstream organizations, some with ties to national movement organizations, emergent organizations campaigning against fracking, radical collectives dedicated to environmental justice, and groups trying to create sustainable communities. In my study, some groups and campaigns tried to bring together many of these different types of grassroots activists, and individuals often drifted in and out of organizations, participating in various campaigns and collective actions.

My account of grassroots activism in Pittsburgh shows how mobilization of the local environmental community ebbs and flows. Four of the five organizations that I observed closely stopped meeting during the period of my observation, but local environmentalism persisted. When campaigns end and when organizations disband or become inactive, the grassroots movement community may become less visible but continues to exist through networks and relationships, individual entrepreneurs, and cultural activities. Some movement organizations also survive, and new ones form in response to emerging and ongoing environmental concerns. Issues and values motivate grassroots activists, and climate change and other problems have only become more urgent over time. Environmental entrepreneurs and organizations attract many supporters persuaded by their framing of these issues, but movement organizers face two pressing needs as they try to mobilize existing and potential members of the movement community. First, they must devise strategies and tactics that will engage activists and build the movement. Second, they have to create some type of organizational structure and culture through which ongoing actions and campaigns can be planned and new activists recruited and integrated into the movement. My study not only demonstrates the importance of strategies and tactics, as well as issues and threats, to mobilization but also reveals the difficulty involved in creating ongoing campaigns that maintain momentum and organizational structures that enhance strategic capacity.

In this introductory chapter, I begin with a brief history of the contemporary American environmental movement, including the roles of both national organizations and grassroots activists in the movement. I then describe my study of grassroots environmentalism in Pittsburgh, where activists have organized around issues of national and international concern, such as fracking and climate change. The study provides a close-up look at a range of grassroots environmental groups and the challenges that they face in mobilizing participants, structuring their organizations, devising strategies and tactics, and building a movement that can make a difference. The Pittsburgh groups that I observed are not representative of any population of grassroots environmental groups, and I do not claim that the experiences of the groups in my study are necessarily typical of other participatory environmental groups. Rather, my goal is to show how grassroots groups with different structures and cultures are able to accomplish what they do and why they are limited in their capacities, thereby contributing to our theoretical understanding of the dynamics and potential of grassroots movements.

THE CONTEMPORARY AMERICAN ENVIRONMENTAL MOVEMENT

The movement to protect the environment predates the 1960s and is not confined to the United States. Numerous scholars have provided histories of the movement, which, in North America, emerged in the late nineteenth and early twentieth centuries as a conservation movement (Bosso 2005; Brulle 2000; Gottlieb 2005; Hays 1979; Kline 2011; Shabecoff 2003; Taylor 2016). The first environmental groups focused on the conservation of public lands and protection of endangered species. Many early conservationists were wealthy sportsmen, who embraced a frontier "manliness" and sought to ensure that game would be available for recreational hunting, which was limited by race, class, and gender (Taylor 2016). The movement expanded its goals and constituency over time, but mainstream environmental groups have remained largely white and middle class (Taylor 2014). Depictions of wilderness and green spaces have focused largely on the experiences of Euro-Americans, making the environment seem like a largely "white space," in which African Americans and other people of color are excluded (Finney 2014:27). Environmental justice activists concerned about the "disproportionate environmental burden borne by African Americans" have challenged this lack of diversity in the environmental movement (Bullard and Wright 1992:40).

A number of organizations that emerged during the Progressive Era and the interwar period, including the Sierra Club, the National Audubon Society, the National Parks Conservation Association, the Izaak Walton League, and the Wilderness Society, experienced remarkable longevity. Some of these early organizations have expanded their agendas greatly and remain active today. After World War I, as Bosso (2005) describes, more groups originated in Washington, DC as a result of federal action on conservation, and groups such as the Sierra Club and the Audubon Society organized outings and focused on recreational activities such as hikes and bird watching. Movement expansion following World War II was fueled by concerns about pollution associated with industrialization as well as continued enthusiasm for outdoor recreation. The ongoing movement and long-time organizations provided a foundation for large-scale growth of environmentalism during the protest cycle of the 1960s (Bosso 2005:45).

The environmental movement flourished as new organizations formed and long-time organizations increased their memberships and expanded their missions in the 1960s and early 1970s. National environmental organizations developed formalized structures that allowed them to engage in lobbying and litigation, while many new groups were formed at the local level in response to federal laws that promoted public participation and court rulings that granted "standing" to citizens concerned about pollution (Bosso 2005; Longhurst 2010). Beyond conservation issues, newer concerns such as the effects of pesticides on the environment, an alarm raised by Rachel Carson (1962) in *Silent Spring*, became important in the 1960s (Dunlap and Mertig 1992:2). In addition to broadening their concerns, local activists also employed new tactics promoted by other movements of the 1960s, such as a "soot in" in New York to protest pollution, where participants sprayed black mist and distributed darkened flowers at the Consolidated Edison building (Rome 2003:544). Antiwar activists drew attention to the devastating use of napalm and herbicides during the Vietnam War, and New Leftists promoted a variety of environmental projects such as community recycling centers (Gottlieb 2005:138). The first Earth Day, on April 22, 1970, demonstrated the power of the modern environmental movement, as huge demonstrations and "teach-ins" modeled on anti–Vietnam War events were held across the country (Rome 2013; Sale 1993).

The environmental movement that emerged in the 1960s included counterculture initiatives in local communities and local citizen groups concerned with air and water pollution as well as new national organizations. As the New Left declined, however, so did "its ability to influence the environmental movement at an organizational level" (Gottlieb 2005:139). The size and number of national environmental organizations, on the other hand, expanded in the 1970s. Many organizations began to professionalize, becoming institutionalized lobbying organizations in Washington. With the founding of the Environmental Protection Agency (EPA) in 1970 and the passage of major environmental legislation, including the Clean Air Act of 1970, the Clean Water Act of 1972, and the Endangered Species Act of 1973, national environmental organizations became involved in intense lobbying and litigation to support legislative initiatives and implement new policies. As Mitchell (1989) shows, the "conservation movement" became the "environmental movement" with the expansion of issues addressed by long-standing organizations and the founding of influential new environmental groups. These included the Environmental Defense Fund, which took up Rachel Carson's call to ban the pesticide DDT, and the National Resources Defense Council, which was supported by the Ford Foundation (Mitchell 1989:88–9). National environmental organizations became professionalized and needed full-time staff with environmental and management expertise to deal with legislative issues and organizational expansion (Mitchell 1989:104–5). Gottlieb (2005:182) describes the growth of "mainstream" environmental organizations and "a revolving door between staff positions in the mainstream groups and government and industry." Bosso (2005) chronicles the growth of an "environmental advocacy community" in the 1970s consisting of organizations filling various policy niches. In the 1980s, after the election of President Ronald Reagan, environmentalists suddenly became outsiders to an unfriendly administration (Bosso 2005:87), but national organizations also expanded greatly as constituents responded to the threats with financial support (Mitchell 1989; Mitchell et al. 1992). In 1990, celebrations of the twentieth anniversary of Earth Day in 140 countries attracted some 200 million people (Gottlieb 2005:262).

National environmental organizations came in for criticism insofar as they were seen as overly institutionalized and coopted by government and industry elites. Writing at the end of the 1980s, Mitchell (1989:107) concluded that the national environmental lobby worked with "reasonable effectiveness in the face of strong opposition from traditionally powerful business and industrial interests," but warned of "the dangers of routinization in advocacy, careerism on the part of staff members, and passivity on the part of the volunteers" associated with professionalization. Critics later dubbed mainstream national environmental organizations "big green" and offered harsher assessments of their effectiveness. In the 1990s, after the election of President Bill Clinton and Vice President Al Gore, a committed environmentalist, threats to the environment seemed to subside and national organizations began to lose members and lay off staff. In his influential book on the decline of the movement, *Losing Ground*, Dowie (1995) noted various reasons for the decline of large environmental groups, including their lack of effectiveness and failure to work with grassroots organizations. In a provocative essay entitled "The Death of Environmentalism," Shellenberger and Nordhaus (2004) added to the criticism of mainstream environmental organizations as having failed to go beyond lobbying for limited legislative proposals to effectively address major issues such as global warming. Increasingly, critics called for revitalization of the movement with a broader range of strategies and tactics and more grassroots action.

While large professionalized organizations have dominated the national environmental movement since the 1970s, the movement always included grassroots components. A number of citizen action organizations formed in the late 1960s and early 1970s in response to opportunities created by legislation that called for citizen participation as well as threats to clean air and water, land, and public health (Ferguson 2015; Gottlieb 2005; Johnson and Frickel 2011; Longhurst 2010). All of the major environmental laws of the 1970s required public notice and participation (Ferguson 2015:8), and citizen activist groups took advantage of these opportunities. Longhurst (2010) describes how citizens in Pittsburgh and other cities across the country became involved in the new wave of citizen environmentalism with the formation of local movement organizations such as the Group Against Smog and Pollution (GASP), which was formed in Pittsburgh in 1969 to address air pollution and remains active today. Ferguson (2015) examines three cases representative of the wide variety of grassroots groups in the American environmental movement: a Montana group of farmers and ranchers opposed to strip mining of coal, a middle-class, largely urban Arizona group concerned with the threats posed by dams to the Grand Canyon and other sites, and a Tennessee group of working-class people who began out of concern about strip mining and evolved into an environmental justice group. As Gottlieb (2005) demonstrates in his history of the American environmental movement, the movement is extremely diverse and complex, consisting of many grassroots campaigns as well as mainstream national organizations.

Influential grassroots mobilizations included a toxic waste movement that emerged in the late 1970s in response to the Love Canal disaster (Szasz 1994). Other toxic waste sites, such as a landfill in Warren County, became the focus of protest led by African Americans, including civil rights leaders who worked to build alliances with poor white residents (McGurty 2007). An expanded anti–nuclear power movement also emerged in response to the 1979 Three Mile Island nuclear accident (Walsh 1988). Local activists opposed hazardous waste landfills and incinerators as well as nuclear facilities near their homes and workplaces (Walsh et al. 1997). Critics labeled such groups NIMBYs (not in my back yard), and some began with limited goals of preventing what came to be called LULUs (locally unwanted land uses). However, many groups expanded their consciousness and began to address a broad range of issues and to adopt a "not in anyone's back yard" (NIABY) perspective (Freudenberg and Steinsapir 1992; Gottlieb 2005; Pellow 2007; Szasz 1994). In some communities, "citizen-worker" groups organized around LULUs and issues such as the destruction of wetlands (Gould et al. 1996).

Many grassroots activists became part of an environmental justice movement that grew rapidly in the 1980s to address the racial, ethnic, and class inequalities associated with a variety of environmental issues (Bell 2013, 2016; McGurty 2007; Mohai et al. 2009; Szasz 1994). Environmental justice groups challenged mainstream environmental organizations to expand their agendas from issues such as the conservation of wilderness areas to the class and racial injustices associated with environmental hazards (Bell 2016; Bullard 1990, 1993; Finney 2014). The toxic waste movement and other community-based environmental campaigns attracted a broader constituency than mainstream national groups, bringing working-class residents and people of color into the movement. Grassroots groups typically focused on the impacts on human health associated with issues such as the siting of toxic waste dumps, which directly affected people in their workplaces and communities. The concept of "environmental racism" began to be used to connect environmental issues with issues of social and racial justice (Bullard 1993; Cole and Foster 2001; Freudenberg and Steinsapir 1992). African Americans became involved in protesting the siting of highly toxic materials in many communities, including those in the southern United States, and civil rights leaders became involved in protests against toxic waste and other environmental hazards that disproportionately affected black communities (Bullard 1990; Bullard and Wright 1992; McGurty 2007). National organizations, including the National Toxics Campaign and the Citizen's Clearinghouse for Hazardous Waste, were formed to work with grassroots groups fighting toxic waste in their communities (Freudenberg and Steinsapir 1992:29).

Radical environmentalists formed grassroots groups such as Earth First! and the Rainforest Action Network (RAN) in the 1980s to engage in militant tactics and promote the "deep ecology" perspective, which was first put forward in the early 1970s as a critique of industrial society and its anthropocentric bias (Devall 1992; Woodhouse 2018). Radical groups such as the Earth Liberation Front and the Animal Liberation Front proliferated in the 1980s and 1990s as a result of increasing environmental threats, frustration with the mainstream movement, and the influences of other social justice movements (Pellow 2014). In the 1990s, a new wave of grassroots environmentalism emerged in the form of biodiversity groups working to protect wilderness and wildlife through litigation and other tactics (Bevington 2009). New Earth First! groups also formed as part of the environmental justice movement. In 2000, a coalition called Rising Tide North America was organized to bring together radical grassroots groups around climate change and other environmental problems (Brulle 2015). These and other groups, including 350.org and its local affiliates, became part of a growing fossil fuel resistance movement (McKibben 2013). The climate change movement includes a wide variety of organizations using different frames and tactics, including mainstream national organizations, new groups formed to address the crisis, and radical groups (Brulle 2015). Radical environmentalists promoted "climate justice" in recognition of the greater impacts of climate change on people in poorer communities and countries who are the least responsible for causing the crisis. They stressed the need for a "just transition" away from fossil fuel dependency that would not put the burden on the most disadvantaged and vulnerable people. In recent years, the climate justice movement has become an increasingly global movement involving many grassroots activists as well as national and international environmental organizations (Mohai et al. 2009; Schlosberg 2007; Spears 2020).

As part of the fossil fuel resistance movement, grassroots activists have engaged in a variety of tactics such as demonstrations and divestment campaigns, and they have mobilized on a number of fronts. In the coal fields of Central Appalachia, local residents initiated the struggle against strip mining, including mountaintop removal coal mining, a highly destructive form of surface mining that expanded in the 1990s (Bell 2016). Groups such as Mountain Justice joined the fight against mountaintop removal, organizing young people to travel to Appalachia and support local residents with annual events such as Mountain Justice Spring Break and Mountain Justice Summer Camp. In Pennsylvania, the Center for Coalfield Justice (CCJ) has battled against destructive coal mining practices, including longwall mining, which can create extensive environmental damages such as the destruction of water supplies. Under the leadership of Veronica Coptis, the CCJ joined with the Sierra Club in filing a lawsuit against a mining company for damages to a state park (Griswold 2017). Grassroots activists have also targeted the oil and gas industries for their construction of pipelines such as the Keystone XL pipeline and the Dakota Access pipeline, mobilizing a large number of activists to engage in civil disobedience and other demonstrations. The anti-fracking movement has organized grassroots actions across the country to publicize the dangers of fracking and to support bans on the practice (Ladd 2018).

Numerous studies point to the accomplishments and promise of grassroots activism, including its impact on social consciousness, public policy, and movement mobilization and coalitions. The massive Earth Day demonstrations in 1970 had multiple, long-lasting effects, including the growth of grassroots movements, media coverage, the stimulation of educational programs in schools, and the strengthening of national political efforts (Rome 2013). The toxic waste movement raised public awareness of the issue, created media coverage, forced the cleanup of some sites, and prompted corporations to consider the environmental impacts of their actions, among other results (Freudenberg and Steinsapir 1992:33–5; Szasz 1994). Movements against practices such as mountaintop removal and fracking have raised public consciousness and resulted in a number of municipal and state bans as well as other legislative actions (Ladd 2018). In framing issues and devising tactics, grassroots groups affect the growth of the movement. For example, by focusing on health issues, some grassroots groups were able to frame issues in ways that allowed them to form coalitions with labor unions (Mayer 2009).

In short, grassroots environmental groups have played a key role in the history of the American environmental movement. Large national organizations have also been important to the movement, and one key question is the extent to which professionalized and grassroots groups can work together or support one another. While grassroots groups are often seen as the solution to an overly institutionalized national movement, few studies examine in detail the challenges and accomplishments of such groups over time. This book assesses the promise and limitations of a sample of grassroots environmental groups in Pittsburgh. Although the study focuses on activism in one city, trends in the local movement reflect developments in national and international environmentalism, such as fossil fuel resistance. My study of the Pittsburgh movement provides a close-up look at the nature of grassroots environmentalism and its potential contributions to the larger environmental movement.

STUDYING ENVIRONMENTALISM IN PITTSBURGH

Once known as the "smoky city" owing to pollution from its formerly vibrant steel industry and the burning of bituminous coal, Pittsburgh has a long history of environmental organizing to mitigate the effects of air and water pollution in the city (Longhurst 2010; Tarr 2003). This industrial legacy, along with long-term environmental problems related to coal mining and more recent problems arising from shale gas drilling in Pennsylvania, makes Pittsburgh a good place to study environmentalism. The city has a rich history of environmental activism: Pittsburgh-area author Rachel Carson raised public consciousness about unseen pollutants with her best-selling book, *Silent Spring*, published in 1962. The 1960s and 1970s were an "environmental era" of movement expansion that generated many new organizations as well as growth in existing ones such as the Sierra Club (Bosso 2005). GASP was part of the wave of citizen activist movements of the period, and the Allegheny Group of the Sierra Club was founded in 1970. More recently, environmental groups have organized to work on issues such as the dangers associated with hydraulic fracturing for shale gas, and Pittsburgh was the first American city to ban fracking. Environmentalists in Pittsburgh have advocated for numerous sustainable practices and changes in the infrastructure needed to make the city more resilient in the face of climate change (DeMarco 2017).

My study of local movement organizations shows how grassroots groups have mobilized activists and carried out collective actions and how they struggle to maintain ongoing organizations and campaigns. On the one hand, it is remarkable how much these groups have accomplished as far as building a movement, raising public consciousness, and, in some instances, influencing policies. On the other hand, grassroots groups face many obstacles, both from external opponents and from internal organizational difficulties, and many decline after a few years. For example, Marcellus Protest helped to kick-start the anti-fracking movement with an exciting protest march in 2010 and played an important role in increasing public awareness and spurring organization in Pennsylvania. But the group had a difficult time creating an effective leadership team and lasting organizational structure and stopped meeting after three years. The Shadbush Environmental Justice Collective was amazingly active and dedicated for a small group, but had trouble expanding and disbanded after four years. The Allegheny Group of the Sierra Club is a much more stable organization that is still active today, but the group also had trouble at times recruiting enough activists to run new campaigns.

One of the difficulties faced by all grassroots organizations, which rely largely on volunteers, is simple burnout; most movement supporters cannot engage in nonstop activities, and many eventually have to divert their attention to families and jobs. Other difficulties faced by grassroots environmental movements are due in part to their strengths. Diverse activists and commitments to democratic practice represent important resources for grassroots movements, but they also present challenges because participants with different backgrounds and ideologies lack a shared understanding of how to organize, even when they have common goals. It is particularly hard to create effective organizations and strategies without strong leadership teams, but it is difficult to develop such teams without agreements about structure and culture. Looking closely at how local groups attract participants and create strategies and tactics, I highlight the internal dynamics involved in building a movement. External threats and pressing environmental problems motivate participants, but groups need workable organizational structures and leadership teams to maintain momentum. They have to frame issues persuasively and provide ways for diverse constituents to become involved in the movement. They need to create alliances and campaigns that will advance movement goals. Strategies and tactics are extremely important to the mobilization and effectiveness of movements, but activists encounter numerous problems in creating organizations with strategic capacity. My research shows how hard it is for groups to continue to devise strategies and tactics and why that is the case by looking at the actual process of talking about strategy in movement organizations.

OVERVIEW OF THE BOOK

Scholars of social movements and contentious politics have written extensively about the process of mobilizing movements and engaging in collective action. Chapter 2 reviews existing work on two key components of movement mobilization, organization and strategy, which are the main topics of subsequent chapters. The chapter discusses work on the challenges of mobilizing, including developing leadership teams, engaging activists, dealing with diversity, and creating organizational structures and cultures. It also examines how movements come up with innovative strategies and tactics, the importance of movement campaigns and communities, and the framing of issues. Lessons drawn from this literature are employed in later chapters to analyze the challenges facing grassroots environmentalists. (Readers uninterested in detailed theoretical discussion can skip Chapter 2.)

Chapter 3 describes the five organizations in my study in detail, providing narratives of their origins and activities. These groups reflect some important trends in the environmental movement, including efforts to create sustainable communities, mainstream organizations, environmental justice groups, and organizing around issues such as shale gas drilling. I draw on the narratives presented in Chapter 3 in subsequent chapters.

Chapter 4 begins to explain the process of mobilization in the local movement by describing the burning environmental issues that motivated many local residents to become active. It also looks at how groups worked to frame issues in ways that captured the experiences and grievances of potential constituents. Chapter 5 then looks at how individuals found their way into movements and organizations and at the role of tactics in mobilizing them. Chapter 6 traces changes over time during the period of my study, looking at how the groups evolved as they created and adjusted their organizational structures and at how networks and relationships in the movement community sustained participation.

In Chapters 7 and 8, I go inside the sample movement organizations, where organizational structures and relationships shape movement participation and strategic decision-making. Chapter 7 looks at the different types of meetings, where activists interact and plan movement activities, demonstrating that meetings are most exciting when they are connected to actions or when they convey stories and information that make activists hopeful that they can create change. Chapter 8 looks at the process of devising strategies and the impacts of leadership and organizational cultures and structures on the ability to create ongoing actions and campaigns.

The conclusion, Chapter 9, brings together the arguments of the previous chapters, summarizing the book's contributions to the study to social movements and its practical lessons for the environmental movement and social movements more generally. In the Appendix on fieldwork methods, I discuss my participant observation and in-depth interviewing, and related methodological and ethical issues.

2

Organization and Strategy in Social Movements

Social movements around the world have demonstrated the power of collective action to bring about significant cultural and political change (Tarrow 2011). The potential for a social movement exists when people have grievances or concerns about a problem and want change; emotions such as anger, resentment, and fear are powerful motivations for collective action. Threats, including environmental hazards, can intensify feelings of grievance and increase incentives for collective action (Almeida 2019). But movements do not simply "happen" in response to upsetting conditions. Climate change, notably, has provoked many complex emotions, including fear, helplessness, and guilt, which have made it hard for many people to respond to the crisis (Norgaard 2011:80). Movements can provide ways for concerned people to act together, and experiences with collective action can create hope that change is possible.

To build movements, groups and individuals have to take the initiative to act on their concerns and to persuade others that their grievances are important and that collective action will be worthwhile. This is easier when emotionally charged issues and events create threats and generate outrage, and movement emergence and growth may be relatively spontaneous in such circumstances. However, both structure and spontaneity play a role in social movements (Killian 1984; Rosenthal and Schwartz 1989; Snow and Moss 2014), and organizers rely on existing and emergent "mobilizing structures" to attract people concerned about movement issues (McCarthy 1996). Movements thrive as supporters become engaged in collective action through movement organizations and communities, and as relationships and networks are forged among groups and individuals who share movement goals.

Mobilization is "the process by which a group goes from being a passive collection of individuals to an active participant in public life" (Tilly 1978:69). Or, if we think of movements as sets of opinions and beliefs within a population (McCarthy and Zald 1977), mobilization is the process of organizing those preferences and gathering resources to support collective campaigns. Both organization and strategy are needed to mobilize a movement. Organization includes networks of participants as well as formal or informal movement organizations. Strategy is a plan of action that involves choices of collective action frames, targets, goals, and tactics. Tactics are forms of action, such as demonstrations, that are frequently aimed at external targets, such as industry and government, but tactics also mobilize supporters. Ongoing actions and campaigns are necessary for movement growth and success, but their creation is no easy feat. Moreover, collective actors do not automatically arrive at the most effective strategies and tactics for achieving their goals. They draw on familiar tactics, which do not always produce gains. Groups do adapt their strategies in response to ongoing experiences, relationships with allies, and interactions with targets, but some movements are more successful than others, due to both their own strategic decisions and the responses of supporters and opponents (Doherty and Hayes 2019). Organizational structures are related to strategy in that some structures, such as effective "leadership teams," lead to greater "strategic capacity" (Ganz 2000).

Movements have the potential to bring about enormous social change, but the work of building a movement and carrying out collective action can be extremely difficult. To understand the process of mobilization, we need close-up descriptions of how groups organize, maintain momentum, and engage in collective action over time. This book details the challenges and opportunities faced by grassroots environmentalists as they tried to build a movement and engage in collective action. It attempts to explain the process of mobilization, broadly understood, demonstrating why movement work is hard and how movements are able to succeed in different ways. In this chapter, I review some of the important lessons that social movement studies offer for understanding how movements organize and formulate strategies. In subsequent chapters, I draw on these findings to understand the efforts of grassroots environmentalists. I begin with a discussion of the process of organizing a movement and then turn to the problem of strategic choice.

ORGANIZING A MOVEMENT

Organization is a key ingredient of mobilization (Gamson 1990; Tilly 1978; Zald and McCarthy 1987), though not all forms of organization are equally effective. Piven and Cloward (1977) warned that the construction of bureaucratic organizations can distract from collective action and, in their study of poor people's movements, found that this form of organization led to declines in contentious action. Other scholars have debated the merits of decentralized, informal, and participatory forms of organization compared to more centralized, formalized, and hierarchical forms (e.g., Leach 2016; Lichterman 1996; Polletta 2002; Staggenborg 1989). Nevertheless, most social movement theorists agree that some sort of organization is needed to attract participants and build a movement, and that it is useful to assess the effectiveness of different types of structures and cultures. In the following sections, I discuss research on differences in several important aspects of organizational structures and cultures that affect mobilization.

Leadership Teams

A number of key processes influence how movement organizations develop structures and cultures that enable them to maintain themselves and formulate strategies. One is the development of leadership teams consisting of organizers who regularly discuss strategy and teach skills to new activists. Ganz (2000, 2009) argues that movement organizations have greater strategic capacity – meaning that they come up with more effective strategies and have a wider repertoire of strategic options – when they provide forums for creative deliberation and develop formal or informal leadership teams that devise and debate strategies. Collective identity, and a commitment to movement values, helps motivate leaders, and work in teams provides feedback on strategic ideas (Ganz 2004:185). Ganz (2000:1016) finds that "organizational structures that afford leaders venues for regular, open and authoritative deliberation; draw resources from a diversity of salient constituencies; and hold leaders accountable to those constituencies – and to each other – are more likely to generate effective strategy than those that do not." Thus, diversity and teamwork are important sources of innovation, and ties among constituents and leaders generate accountability.

Polletta (2002) is similarly interested in the sources of innovation, leadership teams, and accountability. Based on her study of participatory democratic organizations, which attempt to treat members equally and typically make decisions by consensus, she challenges the widespread view that such groups are laudable but ultimately impractical experiments. Polletta identifies several strengths of this form of organizing and shows how it can be strategically useful. One benefit is that participation in decision-making enhances feelings of commitment and solidarity with the group. Another is that the process generates ideas, leading to tactical innovation. There are also "developmental benefits" of participatory democracy, meaning that individuals learn to present arguments, participate in leadership, and engage in collective action. Polletta (2002:11) argues that this form of organizing "is a way to craft new mechanisms of accountability and new norms of leadership." Thus, participatory democracy can potentially create the types of leadership teams and deliberations that Ganz finds critical to effective organizations.

Other scholars have also examined the types of organizational structures that develop leadership and help groups remain mobilized and engaged in collective action. A comparison of patterns of mobilization in four chapters of the National Organization for Women (NOW) identifies organizational characteristics that facilitated ongoing mobilization in two of the chapters (Reger and Staggenborg 2006). These included continuity of leadership as well as leadership teams; chapters where former presidents continued to be active and served as advisors to new presidents enjoyed considerable success in maintaining themselves. Leadership development depended on features such as multiple committees, offices, and staff positions, as well as regular activities such as weekly volunteer sessions. Structures that promoted ongoing interactions, as Ganz (2000) found, were important in allowing leadership teams to exchange ideas and enabling new members to learn from experienced ones. Moreover, mobilizing tactics such as program meetings and workshops, as well as enjoyable meetings and ways for newcomers to get involved easily, were important to ongoing mobilization.

Leadership in Environmental Organizations

This study adds to research that has looked specifically at leadership in environmental organizations by going inside grassroots organizations to show how leadership influences them and what difficulties they encounter. In an extensive study of leadership development in the Sierra Club begun in 2003, Ganz and his colleagues demonstrate both the importance of leadership teams for organizational effectiveness and the difficulties that some regional groups and state chapters have had in developing leadership (Andrews et al. 2005, 2010; Baggetta et al. 2012, 2013; Ganz and Wageman 2008; Han et al. 2011). In comparing different chapters and groups, the researchers found that the most successful ones were those with more committed activists and interdependent leadership teams. However, they found that most of the chapters and groups did little to recruit leaders and to teach Executive Committee (ExCom) members the skills necessary for effective leadership (Andrews et al. 2005). As a result of these findings, Ganz and Wageman (2008) developed a series of workshops that were attended by representatives of four state chapters and twenty regional groups to teach participants how to create interdependent leadership teams and to build relationships and devise creative strategies.[1]

This type of training is important because leadership matters to organizational effectiveness. Using data from their comparative study of Sierra Club chapters and groups, Ganz and his colleagues assessed the influence of available resources, political context, and organizational characteristics and practices on multiple measures of organizational effectiveness, including public recognition, member engagement, and leader development. They found that local political context and resources are important but that "available resources and context must be acted upon by leaders" (Andrews et al. 2010:1227). Regardless of resources and context, organizations with more skilled and committed leaders have a more visible presence in the community (Han et al. 2011). Baggetta et al.

[1] Members of the research team later became involved in the Obama campaign, which employed community organizing techniques to build relationships and grassroots leadership among volunteers (McKenna and Han 2014).

(2013) find that leaders who are highly committed to their associations have an important impact on organizational outcomes. They show that both individual characteristics, such as skills, motivations, and available time, and organizational characteristics, such as the presence of interdependent teams, shared work, and less time spent at meetings, make leaders more committed.

Using data from their survey of leaders serving on Sierra Club group and chapter ExComs, Baggetta et al. (2012) report that ExCom members spent much of their time performing administrative tasks, such as writing newsletters and keeping up websites, and they spent a lot of time at meetings. However, they spent much less time engaging in community outreach and mobilizing, and their reported skill development reflected the ways in which they spent their time. Interestingly, the authors find that the more time ExCom members said they spent at meetings, the less skill development they reported. In puzzling over this finding, Baggetta et al. (2012:130) note that "meetings can take many different forms, and our results show that the particular type of meetings held in these Sierra Club groups and chapters are not positively related to skill improvement in leaders," and they recommend a closer look at what goes on in the meetings and activities of the Sierra Club and other such organizations. In Chapters 7 and 8, I provide this type of detail on what happens at meetings in connection with leadership development and strategic planning.

Whereas Sierra Club groups and chapters have fairly traditional structures such as elected ExComs that operate within a federated organization, other studies have looked at leadership development and the effects of leadership in different types of environmental organizations. Lichterman (1996) found different approaches to leadership in groups with different cultural styles. A group that was part of the US Greens movement (Lichterman 1996:25) practiced a "personalized politics," in which individual activists were each "a locus of political responsibility and efficacy" (Lichterman 1996:34) and all members were seen as leaders. In a suburban community organization, not all of the core members shared the same kind of personalized political identity as the Greens. The leadership of one member was undisputed, and informal "tutelage relationships" emerged in which newer members learned from more experienced ones (Lichterman 1996:95). In an African-American group, which practiced a "communitarian" style, participants saw themselves as representatives of their community and were comfortable with providing input while allowing leaders to direct the group. Thus, leadership practices vary in groups with different types of constituents and organizational cultures.

Activist Engagement and Styles of Participation

Organizers in local groups make different choices in their efforts to engage activists. In *How Organizations Develop Activists*, Han (2014:8) distinguishes between "mobilizing," which involves "maximizing the number of people involved," and "organizing," which means cultivating the skills and motivations of people who are mobilized so that they are transformed into activists and leaders. Han (2014:93) also employs a useful distinction between "transactional outcomes," such as the numbers of people turned out for an event, and "transformational outcomes," which involve "the way the act of participation changes individuals, groups, organizations, and constituencies." In *Groundbreakers*, McKenna and Han (2014) show how the Obama campaign created neighborhood teams that developed relationships and leadership skills among volunteers. In contrast to previous campaigns that relied heavily on staff, the Obama campaign demonstrated that staff could use community organizing tools, such as house meetings, to produce transformational outcomes among volunteers and in communities.

In comparing high- and low-engagement chapters of national organizations, Han (2014) finds that the most effective groups combine transactional mobilizing with transformational organizing. Whereas low-engagement organizations create low-cost opportunities for volunteer involvement, such as writing letters to legislators, high-engagement groups do not simply recruit people for specific tasks but try to develop their skills and motivations as activists and foster relationships among group members. For example, high-engagement groups frequently combined social and political events to allow relationships to develop, whereas low-engagement groups often focused on activities that people could do alone. High-engagement groups provided regular means of involvement, such as phone banks, and they permitted participants "strategic autonomy" in completing the work. They also created "campaign-based teams" that allowed activists to work on projects with clear goals. High-engagement chapters provided training and emotional support that enhanced the capacities of participants (Han 2014:101–7).

The structures of movement organizations are also related to styles of participation. Lichterman (2005:58) notes that a "plug-in" style of volunteering, like the limited tasks that activists perform in Han's low-engagement groups, has become common in American civic and service groups. For example, volunteers often perform circumscribed tasks such as tutoring children or serving meals. In his study of church groups, Lichterman found that this type of volunteering did not require participants to form relationships and did not result in the lasting connections to diverse communities that some groups sought. As Han (2014:90–1) argues, relationship building among activists is critical to ongoing participation in organizations and collective action. In a study of trajectories of grassroots activism over time, Han et al. (2017) find that it is difficult for grassroots groups to sustain volunteer activism over time; in an environmental group they studied, only a small core of activists remained active over time. While attrition rates were high for both online and offline activism, offline activism was somewhat more persistent (Han et al. 2017:38). This suggests that relationships formed among core activists maintain participation, as Chapter 5 shows. While grassroots movements survive because of the "individual enterprise" of activists who put energy into organizing collective action (McCarthy and Wolfson 1996), most activists work within organizational structures. Lasting movements require structures that encourage the formation of relationships and the development of skills and motivations necessary for ongoing mobilization.

The internal cultures of movement organizations, including emotional cultures, affect their ability to mobilize and maintain commitments. In the battle against AIDS, emotions such as anger played a powerful role in mobilizing the activist group AIDS Coalition to Unleash Power (ACT UP), while despair contributed to its decline (Gould 2009). In her study comparing a Catholic Worker community group and an anti–death penalty group called STOP, Summers-Effler (2010) found that the two groups had very different "emotional rhythms," which affected the commitments of volunteers. Whereas STOP was a much less emotionally intense group with volunteers who came and went, the Catholic Worker community elicited strong emotions, such as excitement and anger, and secured attachments from participants, who tended to become heavily involved for many years.

Organizational structures and cultures influence individual trajectories of participation in social movements. Corrigall-Brown (2012) found that many individuals engaged in movement activity episodically rather than continuously over the life cycle. While individual-level factors such as ideology and family commitments affected participation, organizational and relational contexts were also critical influences on identities and social ties within movements (Corrigall-Brown 2012:81). Groups that fostered high levels of interaction and participation in rituals were likely to create ties that led to persistent activism (Corrigall-Brown 2012:93). Persistent activists typically bond to their organizations and leaders (Downton and Wehr 1997:143). As Chapter 6 reveals, relationships among local activists keep grassroots movements alive even when individual organizations decline.

Movement organizations differ with regard to features that affect mobilization, such as the number and nature of group meetings, the types of "task committees" or "working groups" that are formed, the ways in which meetings are conducted, the extent of centralization and bureaucratization, and the number of participants who are actively involved in the group (Gamson 1990; Han 2014; Lichterman 1996; McCarthy and Wolfson 1996; Staggenborg 1989). Some groups create "forums" or interactional spaces that allow participants to have critical discussions about group identity, process, and strategy (Lichterman 1999). Grassroots groups can be organized in ways that promote the exchange of ideas and allow participants to learn from their experiences. Polletta (2002:210) says that when she asked her informants "to describe a good meeting, they usually described one percolating with creative ideas." At their best, grassroots organizations generate innovative tactics and allow participants to reflect on how well their structures and actions are working.

However, not all groups develop procedures for internal reflection, and some fail to recognize structural problems and learn lessons from organizational and tactical mistakes. Freeman (1972) famously warned of "the tyranny of structurelessness" in women's liberation groups that tried to disavow leaders and structures, but ended up developing informal structures and unaccountable leaders based on friendship networks. In her study of newly emerging groups, Blee (2012) found that leadership was rarely discussed, either in progressive or conservative groups. While activists wanted to avoid the control and power associated with leadership, unacknowledged leadership did emerge (Blee 2012:124–6). Just as Lichterman (2005) found that few groups successfully practiced "reflexivity," Blee (2012:99–100) observed that few actually took time for "collective learning" and that most groups did not do much self-education to learn from their mistakes. In her study of the Chicago Women's Liberation Union (CWLU), Strobel (1995:155–6) found that the organization developed some procedures for internal learning, such as a Liberation School course for new members and the practice of publishing lengthy evaluations of work projects in the newsletter. Yet even within this organization, which was highly committed to self-evaluation, Strobel found that the learning of lessons was selective. While the CWLU learned lessons about organizing that were useful to members who participated in subsequent groups, the organization never learned how to handle ideological diversity (Strobel 1995:158–60). In short, the creation of exciting and functional grassroots groups that continue to develop their capacities is extremely challenging – but worthwhile when successful.

Diversity

One of the challenges in mobilizing a movement is to create an organizational culture and structure with features, such as decision-making procedures, that appeal to diverse constituents. Diversity[2] is important because it brings new ideas, resources, legitimacy, and networks to movement organizations. Organizational choices are influenced by the backgrounds and ideologies of activists and by their experiences and interactions in an emergent movement. Activists create the types of organizations with which they are familiar, and different types of structures appeal to different types of activists. A good example of this occurred in the second-wave American women's movement, where activists with backgrounds in traditional voluntary organizations and political parties created bureaucratic movement organizations, while women with experience in the civil rights and student movements set up collectives (Freeman 1975). The different experiences of women continued to affect the women's movement

[2] Although "diversity" is often used to describe racial or ethnic inclusiveness, I use the term more broadly to include differences in movement experience and ideology, class, gender, age, place of residence, and other background variables, as well as racial and ethnic differences. Most of the groups in my sample were not racially diverse, but some were diverse in other ways.

as new generations and cohorts entered the movement with varying perspectives and feminist identities (Whittier 1995, 1997). In addition to having different approaches to leadership, Lichterman (1996) observed that grassroots environmentalists with different backgrounds engaged in different types of community building and formed organizations with different cultures and structures. In the personalized style of organizing, meetings were committed to individual participation and debate as participants engaged in practices such as prolonged "group process" sessions to assess their organizations. In contrast, communitarian activists formed more conventional types of organizations with committees, boards, and general membership meetings (Lichterman 1996:109).

Organizations that manage to attract diverse constituents may enjoy strategic advantages. Ganz (2004:187) argues that heterogeneous leadership teams provide access to greater networks and resources and potentially make better strategic decisions if the team is able to "foster minority expression" and learn by experimenting with various strategic ideas. Polletta (2002:209) similarly points to input from diverse participants as a source of tactical innovation in a decision-making process that is "deliberative and experimental." Moreover, even conflict among diverse participants can lead to creativity and growth in a movement. In her study of "strategic adaptation" in US women's jury movements, McCammon (2012) argues that "diverse memberships" in movements, which may come through participation in coalitions, aid in bringing new information and tactics to groups, and that disagreements can help groups to critically evaluate tactics. Kretschmer (2019), in a study of the National Organization for Women, found that internal diversity and conflict often benefitted the organization and the women's movement by creating change and new organizations, which typically ended up working with NOW. While it can be difficult to create and maintain diverse organizations, outsiders who provide new perspectives and connections to other groups are likely to be an important source of strategic innovation (Ganz 2000, 2004; Polletta 1999, 2002).

Despite its attractions, diversity does not automatically yield benefits. Because newcomers can challenge group ideas about "what members should be like," grassroots groups may fail to welcome newcomers or listen to their ideas (Blee 2012:74–5). Groups that draw in diverse constituents must reconcile cultural differences associated with class, race, gender, and other identities. In reviewing research on how diversity affects mobilization by community organizations, Walker and Stepick (2014) find that diversity is a major challenge to community organizing processes insofar as ascriptive identities such as race or ethnicity prevent the creation of new identities that transcend differences. Successful multiracial groups, for example, need to go beyond bridging ideological differences "to building organizational cultures that are attentive to group differences" (Walker and Stepick 2014:962). Research shows that various types of "bridging" practices help overcome the challenges created by diversity. "Bridging cultural practices" such as prayer and other religious practices helped unite racially and socioeconomically diverse faith-based coalitions (Braunstein et al. 2014). Cross-movement "bridging processes," including a strategic division of labor and the development of mutual commitments, resolved differences in a diverse alliance of environmental justice and peace activists (Beamish and Luebbers 2009). In coalitions among diverse groups and across movements such as the environmental and labor movements, leaders who are "bridge builders" (Rose 2000) or "coalition brokers" (Obach 2004) have played important roles in making such alliances work.

Collective identity is often viewed as a means of overcoming conflicts that may result from diversity. The Coalition of Labor Union Women was able to bring together feminists and labor unionists by framing women's issues as labor issues and creating a collective identity of "labor feminism" (Roth 2003, 2008:215). In contrast, the "ethos of organizing one's own" that became popular during the 1960s and 1970s discouraged coalition work across racial and ethnic lines (Roth 2010). Moreover, not all efforts to create intersectional collective identities succeed. A study of "multi-identity work" in lesbian and gay organizations found that, despite the emphasis on diversity in movement discourse, reliance on corporate funding sources and attention to organizational maintenance perpetuated inequalities (Ward 2008). In the global justice movement, Fominaya (2010) finds that effective and participatory assemblies are important in fostering collective identity among diverse participants. But she also notes that collective identity can be developed at different levels; even if a particular movement group fails to create a cohesive identity, individuals may develop collective identity at the network or movement level.

Diverse groups sometimes come together through movement campaigns and coalitions, which can create solidarity among diverse participants (Meyer and Kimeldorf 2015; Staggenborg and Lecomte 2009). However, not all alliances produce such solidarity, and cultural differences may be a source of tension in building intersectional coalitions. As Lichterman (1995:523) notes, "different practices of community could easily lead to difficulties and misunderstandings in alliance building, even when groups agreed on ideology." In his study of how church groups attempted to reach out to the larger community, Lichterman (2005) shows that the cultural "customs" of groups – ways of speaking and acting – limited their ability to relate to other groups in the community. While these customs are often taken for granted, groups that practice "social reflexivity," which involves talking about their relationships with outsiders, are more likely to be successful in creating bridges to other groups (Lichterman 2005:45). However, Lichterman found that group customs placed limits on such reflexivity, and few groups in his study were able to really talk about the problems they had in relating to outside groups and institutions.

Summary

Research on social movements shows how structural and cultural characteristics affect mobilization and strategic capacity. Leadership teams are particularly important for developing participant skills and ongoing strategies and tactics, but many groups have trouble organizing effective leadership teams. Decision-making structures are also important to organizational commitment and skill development, though various challenges are associated with different types of structures. Committee structures are venues for leadership development and participation, but groups do not always create and maintain them. Mobilizing tactics, such as program meetings, help bring in new activists, but groups must expend energy to employ such tactics and do not always have the resources to do so. Styles of participation vary in different groups, affecting networks, relationships, and tactics, as well as group culture and structure. Diversity is an important source of innovation and resources, but it can also be a source of conflict, creating challenges to building solidarity and collective identity. Table 2.1 summarizes these characteristics of movement organizations and associated advantages and challenges.

MOVEMENT STRATEGIES AND TACTICS

Strategies and tactics are critical both for accomplishing goals and for building movements. Ongoing movements require ongoing collective action; to create a movement, activists and organizations need to devise strategies and tactics that involve participants and produce desired outcomes. Movements that survive over a number of years consist of a series of actions and campaigns that attract participants, media attention, and public support. Collective action strategies are influenced by external political and cultural environments and by internal features of movements and organizations (Meyer and Staggenborg 2012). Both agency and structure matter; activists make strategic decisions that affect outcomes, and they are constrained by large-scale and organizational-level structures.

Scholars have examined the ways in which movement strategies are shaped by ideology, political opportunities, countermovements, organizational structures and resources, the backgrounds and identities of participants, and cultural contexts, and they have examined the ways in which strategies and tactics affect movement outcomes (e.g., Amenta et al. 1999; Andrews 2004; Banaszak 1996; Freeman 1979; Gamson 1990; Luders 2010; McCammon 2012).[3] Collective actions can build movements even if they have no immediate or direct impacts

[3] At the macro level, Tilly (2008) demonstrates how "repertoires" of collective action and "contentious performances" change over time with large-scale changes such as the expansion of nation-states. Historically, collective actors engage in a limited repertoire of familiar actions shaped by their interactions with authorities and other targets. Within an historical context, collective actors respond to a variety of environmental factors, including political opportunities and threats, cultural norms and practices, and the actions of allies and opponents. Rossi (2017:37–8) argues that Tilly's focus on a repertoire of contention that is "public, militant, and glacially slow to change" needs to be supplemented by attention to nonpublic and noncontentious elements of a "repertoire of strategies" that "is mostly defined in its relationship to medium- and short-term changes in the political context."

TABLE 2.1 *Characteristics of movement organizations*

Organizational characteristics	Advantages	Challenges
Leadership teams	• Skill development • Continuity of leadership • Strategic capacity	• Organizers needed to build teams • Burnout • Need for ongoing issue campaigns
Decision-making structure (extent of participatory democracy, formalization, centralization)	• Participatory structure encourages skill development, tactical innovation, solidarity • Formalized structure helps with continuity of leadership, ongoing activities • Centralization may be more efficient	• Length of time to make decisions/take action • Extent of democratic engagement • Inclusiveness of diverse constituents
Committee structure (task and issue committees, working groups, boards, etc.)	• Committees and working groups are venues for development of leaders • Division of labor spreads work around and facilitates multiple campaigns • Committees and working groups provide roles for new participants	• Need to recruit committee leaders • More complicated decision-making process • More administrative work means less time for community outreach and skill development
Mobilizing tactics (program meetings, workshops and forums, community outreach)	• Ways for organization to meet regularly and develop agenda, train volunteers • Leadership development and engagement • Inclusion of participants from outside organization • Community recognition	• Labor and time intensive • Need to combine with strong recruitment practices • Need skilled leaders to effectively engage in outreach
Styles of participation (extent of engagement, formal or informal roles, emotional involvement)	• High engagement can create lasting ties and collective identity • Easy tasks provide "low cost" opportunities for involvement • Exchange of ideas • Emotional commitment • Internal reflection	• Resources needed to develop skills • Difficulties in forming relationships and creating lasting connections • Informal power structures based on friendship networks • Clashes of participatory styles

TABLE 2.1 (*continued*)

Organizational characteristics	Advantages	Challenges
Diversity (movement experience and ideology as well as race, ethnicity, class, gender, age, and other background variables)	• Increased numbers of participants, scope of networks • Increased innovation, resources, legitimacy, and accountability	• Potential conflicts, challenges to building collective identity and solidarity

on targets. To sustain momentum, a movement needs an organizational infrastructure capable of generating strategies and tactics and nourishing activists and relationships in the movement community. Organizers need to frame issues in ways that convince potential supporters of the movement's interpretation of the problem and proposed solutions and motivate people to participate in collective action (Benford and Snow 2000). Movement activists also adjust their strategies as they engage in collective action and learn from their experiences. In the following sections, I discuss the importance of movement campaigns and communities and framing before turning to the importance of strategic innovations and adaptations and the question of why some movements and groups are better at this than others.

Social Movement Campaigns and Communities

Social movements typically consist of a series of campaigns, which are strategically linked episodes of collective action aimed at achieving movement goals (Marwell and Oliver 1984; Tilly 2008; Tilly and Wood 2013). Social movement organizations (SMOs) and activists are part of social movement communities (SMCs), which provide organizational and ideological bases for the emergence and maintenance of SMOs and campaigns (Buechler 1990; Staggenborg 1998; Taylor and Whittier 1992). Movement campaigns are linked to movement communities in that they are supported by the structures, cultures, and resources of SMCs and in turn have important impacts on the networks and organizations of SMCs (Staggenborg and Lecomte 2009). Campaigns influence subsequent campaigns by altering movement networks and political opportunities and by providing models of "contentious performances" (Tilly 2008).

The interactions of activists within larger communities are critical to collective identity and ongoing mobilization. Activists in movement communities form relationships with one another that sustain movements even when organizations decline or environmental conditions for mobilization are difficult. For example, Bell (2013) found that friendships in the activist community of women in the fight for environmental justice in Appalachia were a source of strength that helped them overcome the isolation of being "radicals." The shape of a movement community affects organizing for a new campaign, which can in turn alter the local movement community by creating new discourse, frames, networks, coalitions, and activists (Staggenborg and Lecomte 2009). In some instances, ad hoc groups organize in response to a "triggering incident" in a local community; some draw on "leaders and networks from previous reform movement activities," while other lack such affiliations (Woliver 1993:4). Such groups are typically short-lived, but they can have lasting impacts. Woliver (1993, 1996) found that ad hoc groups with ties to larger movements were more likely to be successful and that after such "sputtering groups" declined, they left behind changes such as new political consciousness, new forms of identity, and networks. In a study of three movement organizations in Vancouver, Carroll and Ratner (2001:608–9) argue that SMOs need to accomplish three core tasks as they create and sustain an "oppositional culture": build community by creating collective identity and connections among members, empower participants by nurturing them and enhancing their capabilities, and engage in collective action. The authors found that providing services was a good way to create community but that there is a tension between serving the needs of constituents and engaging in political action (Carroll and Ratner 2001:612–13).

Collective action is an important source of collective identity and solidarity within movement communities. Whether part of a campaign or a single event, collective action allows more people, beyond those who work to maintain SMOs, to get involved in a movement. Periods of activity in a movement provide ways for supporters who are not normally very active to become involved, building commitment and community in the movement (Downton and Wehr 1991). During protest events, consciousness raising is likely to occur as participants develop solidarity and collective identity, but only "if it is founded on existing collective beliefs" (Klandermans 1992:92). For example, protesters used a blockade tactic and leaders took personal risks during the student divestment movement at Columbia University, resulting in increased commitment and solidarity in a coalition group that spent several years developing movement frames (Hirsch 1990). Participation in the living wage campaign in Chicago had a powerful impact on individuals, creating solidarity with the movement, and the campaign helped bring together diverse constituents as a result of their shared experiences (Meyer and Kimeldorf 2015). Indeed, campaigns can play an important role in creating new alliances and coalitions within movement communities as participants interact with people that they would not ordinarily encounter and build bonds of trust (Levi and Murphy 2006; Staggenborg and Lecomte 2009). Even when the collective action is short-lived, the experience of participation often endures and affects subsequent actions and campaigns. This could be seen after protests of the G20 Summit in Pittsburgh and Toronto, which introduced activists to skills and tactics that were used in subsequent campaigns (Staggenborg 2015; Wood et al. 2017).

Framing

Framing is an important element of social movement strategy, which helps connect participants in movement campaigns and the movement community. The concept refers to the active process of "meaning construction" by organizers and SMOs as they try to explain their concerns and recruit others to the movement (Benford and Snow 2000:614). Theorists have examined different ways in which frames are used in movements to attract participants, create coalitions, and direct collective action tactics. Three "core framing tasks" involved in creating collective action frames are diagnosis of the problem, proposal of solutions, and motivation of supporters to engage in collective action (Benford and Snow 2000:615–18). Empirical studies of collective action framing show that movement activists are strategic actors who try to construct frames that will advance their cause. Framing involves creativity, and not all frames are equally convincing. Frames are more or less persuasive depending on features of the frame itself and on the characteristics of those who construct frames and audiences of potential movement constituents (Noakes and Johnston 2005:12).[4]

Numerous researchers discuss the ways in which framing is influenced by the cultural and political environment (Benford 1993; Donovan 1995; Kubal 1998; McCammon et al. 2004). Political conditions and cultural norms not only place constraints on framing but also provide opportunities for movements to take advantage of popular grievances and concerns. Some movements and SMOs are able to adapt to changes in the environment, while others fail to do so, in part because of the composition of the movement but also as a result of their interactions with the targets of frames (McCammon et al. 2004). Some SMOs have more flexible organizational structures and identities that allow them to adjust their framing and take advantage of opportunities for media coverage

[4] Some frames have more "resonance" than others for various reasons, including their "empirical credibility" or "fit between the framings and events in the world" (Benford and Snow 2000:620). In part, the ability to come up with effective collective action frames and innovative tactics depends on the availability of generic "master frames" that can be adapted by specific movements (Snow and Benford 1992). For example, the "rights frame" associated with the civil rights movement's "goal of integration through nonviolent means" was a particularly flexible frame that influenced the framing and tactics of other social movements such as the gay rights and women's rights movements (Snow and Benford 1992:145). Snow and Benford (1992) argue that movements can create master frames that help generate what Tarrow (1983, 2011) calls a "cycle of protest" or "cycle of contention" consisting of multiple movements during a time period such as the 1960s.

(Rohlinger 2002, 2015). Movement organizations typically engage in both "back region" framing among activists, which may be more radical, and more strategic "front region" framing for public consumption (Kubal 1998).

Framing is important because language influences the reception of movement ideas and because the type of "contentious language" used is related to the forms of action employed (Tarrow 2013). The framing of a problem often suggests certain types of strategies and tactics and rules out others. In addition, framing affects alliances within and across movements. "Frame bridging" is a strategic process, whereby organizers facilitate alliances by connecting different but ideologically compatible frames to one another (Benford and Snow 2000:624). Organizers engaged in frame bridging can create a master frame that connects ideological concerns and unites constituents across movements, such as feminists, peace activists, and unionists, to expand movement campaigns (Carroll and Ratner 1996; Gerhards and Rucht 1992).

Tactical Innovation and Strategic Capacity

One important question is how collective actors come up with innovative tactics that allow a movement to attract supporters and make gains. McAdam (1983) argues that novel tactics are critical to both the growth and success of movements; in explaining the development of the civil rights movement, he demonstrates that activists came up with a series of tactical innovations as targets responded to movement tactics and movement actors in turn devised new strategies. But not all strategies and tactics are innovative or effective; collective actors face numerous constraints and trade-offs as they devise strategies in the context of internal and external interactions with supporters and targets (Jasper 2012). These interactions occur in organizations, at protest events, at meetings with legislators and politicians, during government hearings, in coalitions, in communities, and on the internet. They occur in cultural as well as political realms, and tactics include "cultural performances" that are part of the current repertoire as well as directly political actions (Taylor and Van Dyke 2004). Within movement communities, strategic choices are influenced by activist networks and relationships that are sources of collective identity, resources, and tactical models (Meyer and Staggenborg 2012). And strategies change over time in response to experiences and interactions with allies and targets (Rossi 2017).

The development and diffusion of innovative tactics often occurs through relationships in organizational and community contexts. The 1999 World Trade Organization (WTO) protests in Seattle popularized a set of street tactics that included black blocs, giant puppets, lockbox blockades, and jail solidarity tactics, and the diffusion of these "Seattle tactics" led to a period of tactical experimentation by multi-issue coalitions in other cities (Wood 2012). In the Climate Action Network, a new generation of climate activists interacted with the "old guard" of experienced activists to produce innovative tactics in an "insider/outsider coalition" (Aunio 2012). Collaboration among SMOs facilitates tactical diffusion, and tactical innovation is likely to occur when multiple claims are made at protest events, leading groups to experiment with new tactics (Wang and Soule 2012, 2016). Many innovative tactics are devised on the internet (Bennett and Segerberg 2013; Carty 2015; Earl and Kimport 2011), and new digital technologies influence the development and spread of protest forms (Polletta 2012:47). However, tactical diffusion may fail to occur in the absence of organizational support and resources to gain media attention as well as increased repression; under these conditions, a tactical innovation may decline before it takes off (Galli 2016).

At the organizational level, networks, leadership, resources, and decision-making structures all affect strategic choices, and some types of organizational structures enhance strategic capacity (Ganz 2000, 2004). Informal and decentralized groups may have an advantage in producing innovative tactics (Staggenborg 1989), and participatory democracy aids in brainstorming ideas (Polletta 2002). Resources and leadership enable organizations to remain viable and to engage in a range of effective protest tactics (Cress and Snow 1996). Andrews (2004:22–5) identifies leadership, organizational structure, and resources as components of "movement infrastructure" that affect strategies and tactics and the ability of a movement to effect social change. McCammon (2012) shows how state movements that adapted to changes in their environments by evaluating and adjusting their strategies and tactics were able to win women's right to participate on juries sooner than those that failed to adapt. Factors that made some movements better able to adapt strategically included diverse memberships, bridge leaders who linked different groups together, and the ability to actively assess political opportunities and threats and to learn from past experiences (McCammon 2012:18–33).

Researchers also emphasize cultural influences on strategies, including various cultural constraints as well as the frames that activists use in justifying their actions (Lichterman 1995, 2005; Polletta 2012; Taylor and Van Dyke 2004). Lichterman (2005:55) defines culture as "a set of publicly shared, symbolic patterns that *enable and constrain* what people can say and do together." The larger civic culture affects group customs, such as the vocabularies used in public life and the ways in which roles, such as that of the volunteer, are defined (Lichterman 2005:57–8). Institutional settings, such as courts, have various cultural conventions that activists typically consider in planning their strategies; they may choose to challenge conventions despite the risks in doing so (Polletta 2012:49–51). Movement groups are constrained by the larger culture, but they also develop internal cultures that support collective action. By engaging in various types of protests, activists develop cultures of protest, collective identities, and cultural skills and resources (Taylor and Van Dyke 2004). As Blee (2013) warns, however, grassroots groups can also develop "implicit discursive rules" that close off some tactical options without deliberation. She points out that leaders can have an impact on tactics but that the development of leaders is

TABLE 2.2 *Influences on strategic capacity*

Political and cultural environment

- Interactions with opponents and targets
- Cultural norms and practices
- Political opportunities and threats
- Repertoires of collective action
- Collective action frames
- Media attention

Movement community

- Movement coalitions and alliances
- Organizational infrastructure
- Relationships and interactions among groups and individuals
- Diversity of support
- Internet communications
- Availability of resources
- Bridge leadership
- Impacts of previous collective actions and campaigns (tactical models, skills, networks, collective identity)
- Frames, discourse, and political consciousness

SMO organization and culture

- Leadership teams
- Diversity
- Interactions among members
- Resources
- Structure (decision-making, committees, forums, etc.)
- Organizational learning
- Framing and discourse
- Mobilizing tactics
- Identity and ideology
- Emotional culture

also shaped by these implicit rules, making it unlikely that they will challenge the group's tactical direction (Blee 2013:675).

Summary

Strategies and tactics are critical to social movement mobilization and outcomes. Social movements consist of collective actions and campaigns orchestrated by activists and organizations. Even when organizations decline, activists remain connected through networks and relationships in movement communities. Framing is one process that creates ties among diverse groups of activists who sometimes join together in coalitions and campaigns. Collective campaigns strengthen ties among individuals and groups in movement communities and influence subsequent campaigns and coalitions. Nevertheless, it is difficult for movement actors to continually devise and carry out collective actions, and not all strategic choices are equally effective. Movement organizations and alliances differ in their strategic capacities. Influences on strategic choices include: organizational features such as leadership teams, decision-making structures, committees, resources, and types of participants; interactions within organizations and between individuals and groups in the movement community; and elements of the political and cultural environment. Table 2.2 summarizes key factors affecting strategic capacity at the levels of the political and cultural environment, the movement community, and the movement organization.

CONCLUSION

A number of issues are important to a close-up study of how movements organize and formulate strategies and tactics. One key question is how activists choose to organize themselves, inside and outside of organizations. Much organization is done through email and social media, and individuals are also connected informally through networks and cultural groups within social movement communities. Social movement organizations are also important, and emerging groups need to decide how to structure themselves. Decision-making procedures vary greatly, with some groups operating by consensus and others by parliamentary procedure. Some groups have regular meetings, while others rely heavily on online communications. Meetings may be open or closed, and diversity may be encouraged or discouraged by group practices. Leadership teams may or may not be nourished, and interactions may be extensive or limited, depending on the forums created by the group. Members may be more or less involved, depending on the opportunities created for participation and development of skills. Working groups or committees might be created to provide ways for members to become involved, and they may be more or less successful.

These structures, and the internal culture of the group, have an important impact on strategic deliberations. Some movement organizations are part of federated organizations, which influence their strategic choices. Some groups are more willing than others to participate in coalitions, which may broaden their strategic range. Structural arrangements, and the diversity of groups, affect their ability to innovate strategically, to follow up on their strategic plans, and to assess the effectiveness of strategies and tactics. Strategy is also influenced by interactions with other groups in the movement community and with targets, such as legislators. Movement alliances are often critical to the ability to devise campaigns, which keep movements alive and have various internal and external impacts. Collective action has important impacts on individual participants and on movement groups as well as on targets, and actions and campaigns at one point in time affect subsequent collective action. Strategic choices, including how issues are framed and what tactics are chosen, are critical to movement mobilization and outcomes, but the process of mobilizing a movement and creating effective strategies is extremely challenging and complex. We still know little about the process of strategic decision-making in protest groups, and a number of scholars have called for more in-depth research, using methods such as participant observation, on the ways in which activists develop strategy in interaction with other actors (Jasper 2004, 2012:31; Maney et al. 2012; Polletta 2002:225). In the following chapters, I draw on lessons from social movement research while providing needed detail on the struggles of grassroots environmental groups.

3

Grassroots Environmentalism in Pittsburgh

The Pittsburgh environmental movement community consists of affiliates of state, national, and international organizations as well as free-standing local organizations. These groups address long-standing problems such as air and water pollution as well as more recent concerns about fracking, environmental justice, and climate change. In this chapter, I describe my sampling of the grassroots environmental movement in Pittsburgh. For the five groups that I studied intensively, I provide narratives of their origins and major activities as well as a brief description of their organizational structures and ideological approaches in this chapter (see also Staggenborg 2018; Staggenborg and Ogrodnik 2015). In subsequent chapters, I draw on these cases to look more closely at processes of mobilization, including internal organizational dynamics and strategic decision-making.

SAMPLING THE MOVEMENT

Initially, I spent some time becoming familiar with various organizations and activities in the local environmental movement community before selecting a sample of groups and activities to study intensively. Because I am interested in grassroots mobilization of local participants, I focused on groups with active volunteers rather than ones that were largely staff run. Beyond looking at established organizations, I attended public events and informal gatherings advertised on various email lists and websites. This is important because not all grassroots activism takes place in the context of organizational meetings; movements include informal gatherings and educational forums as well as formal movement organizations. For instance, I attended a showing of a Transition film organized by a small group of people who founded Transition Pittsburgh and later attended informal discussions about creating sustainable communities held by the group in a local café. My exposure to the anti-fracking movement occurred after I viewed the documentary film *Gasland* and then attended a meeting called by an informal collection of local activists concerned about shale gas drilling. The group wanted to call attention to the issue by organizing a large demonstration in Pittsburgh to coincide with a gas industry conference, and I decided to participate in an intense round of organizing for this demonstration. This allowed me to get to know a number of local environmental activists during what I considered an exploratory phase of my research. As an environmentalist myself, I was happy to contribute my time and energy; later, when I sought formal permission for my study, I was able to gain acceptance because a number of participants knew me as someone who had contributed to the organizing effort. (See Fieldwork Appendix for further details.)

After learning the lay of the land in the local environmental movement, I ended up focusing my study on four groups, and later a fifth one that grew out of the anti-fracking movement. I also continued to follow other developments in the local movement by joining email lists, talking to participants, and attending occasional meetings and events such as the launches of Pittsburgh 350.org and the Environmental Justice Committee of the Thomas Merton Center. The groups that I studied intensively were selected on the basis of what was happening in Pittsburgh to create a sample that varied in structure and ideology. While I could not sample all of the different strands of environmental activism in the city, I selected what seemed to be the most active types of groups in addition to choosing ones with different ideologies and structures. The bulk of my data consists of participant observation at the meetings and events of five organizations, supplemented by interviews with some key participants. I attended about 20 meetings and events held by Transition Pittsburgh and gathered 17 interviews with members.[1] In the case of the Allegheny Group (AG) of the Sierra Club (SC), I attended about 70 meetings and events and conducted 8 formal interviews. For Marcellus Protest (MP), I attended about 75 meetings and events, and conducted formal interviews with 16 participants. In the case of the Shadbush Environmental Justice Collective, I attended more than 50 meetings and events and formally interviewed 6 core members of the collective. I attended some 55 meetings and events of Protect Our Parks (POP) and had many informal conversations with activists, but did not conduct any formal interviews.[2]

[1] For this group, my research assistant, Corinne Ogrodnik, conducted 13 of the 17 interviews, and she also attended a couple of events. For the other groups, I conducted all of the interviews and did all of the fieldwork myself.

[2] Because I had so many informal conversations with POP members, and because I had interviewed a number of core participants earlier regarding their roles in other groups, I decided not to conduct formal interviews with them as POP members. In addition to the formal interviews noted, I also conducted short, tape-recorded interviews (about 15 minutes in length) with nine people on the bus ride to Harrisburg to protest at the inauguration of Governor Tom Corbett on January 18, 2011. Most were participants in Marcellus Protest and several were Sierra Club members.

Of these five groups, only one continued to meet regularly by 2017 and several survived largely online – suggesting the importance of online tools in keeping movements alive. Transition Pittsburgh ceased holding organizational meetings in 2013, but continued to exist through the online Meetup tool, which some members used through 2016 to organize gatherings. The Shadbush Environmental Justice Collective disbanded in 2014 to give rise to a new organization called Three Rivers Rising Tide, which was active for a few more years. Marcellus Protest stopped meeting in 2013 but continued to operate through its website and email lists. Many of its members became active in the Protect Our Parks coalition that was formed in 2013, and its meetings became a substitute for MP meetings. POP continued to meet regularly until the summer of 2016, after which the group decided to suspend meetings until new issues or threats arose. The Allegheny Group of the Sierra Club is the one group in my study that continued to hold regular meetings and engage in ongoing activities; indeed, the AG exhibits remarkable longevity after meeting for some fifty years to date.

Two of these groups are part of larger structures: Transition Pittsburgh is a local initiative of an international organization that aims to create local models of sustainable communities. Transition is not structured like a typical movement organization but is more of a network connecting a number of local projects aimed at creating resilient and sustainable communities. While active, Transition Pittsburgh was largely apolitical and encouraged green entrepreneurship and community organizing. Although there was some overlap in membership with other local environmental groups, Transition Pittsburgh attracted a somewhat different constituency, including many young people and a wide range of ages, green entrepreneurs, New Age spiritual types, and some class and racial diversity. The Allegheny Group of the Sierra Club is part of a federated organization with state and national as well as regional levels, and it has a more formalized structure than some other local environmental groups. Because the local group is run by volunteers, I was interested to see how this structure supported organizational maintenance and affected strategies and tactics. Unlike other groups in my sample, the Sierra Club engages in electoral politics as well as a range of other tactics, and it attracts a somewhat "mainstream" constituency.

The other three groups in my sample are local ones that are part of the large movement against shale gas drilling in Pittsburgh and Pennsylvania. Marcellus Protest originated in 2010 to respond to the growing threats posed by fracking. It is interesting as a fairly heterogeneous group, which attracted participants who varied by residence (rural or urban), age, education, and social class (though not race, being largely white), and was a major player in the anti-fracking movement in the city and state. The Shadbush Environmental Justice Collective focused heavily on shale gas drilling, and some of its members also participated in MP. However, Shadbush was organized as a small collective that was much more ideologically and organizationally cohesive than MP.

Shadbush began in opposition to Mountaintop Removal, and the group considered itself a radical environmental group with a multi-issue approach despite its later emphasis on fracking. Protect Our Parks came out of an effort by Shadbush to create a local campaign and became an important coalition in the anti-fracking movement for several years.

Transition Pittsburgh

Transition is an international network of initiatives that began in Ireland and the United Kingdom as an attempt to provide positive local responses to climate change, "peak oil," and the 2008 economic crisis (Hopkins 2008, 2011). Despite the failure of environmental action at national and international levels, the premise of Transition is that it is possible for local communities to create models of how to transition away from reliance on fossil fuels that are actually preferable to consumerist, carbon-intensive living. Transition Town groups in the United Kingdom and elsewhere have engaged in a variety of interesting initiatives such as local currencies, skill sharing, community gardens, permaculture, and energy projects. The model attracts participants with concrete projects and the potential to strengthen community relations. It also raises interesting strategic questions: How do participants expect concrete projects to make a difference with regard to environmental problems? How does the movement go from community projects to political and cultural change? How might these projects go from local efforts to broader "strategic thinking" and "scaling up" (Hopkins 2011:85)?

Transition Pittsburgh was founded by several local activists interested in sustainable communities who learned of Transition either by reading *The Transition Handbook* by founder Rob Hopkins (2008) or through personal contacts. The group's strategies and organizational structure were influenced by the "twelve steps" Transition model, which is laid out in detail by Hopkins (2008) and summarized in two versions of the *Transition Initiatives Primer*, one created by Transition UK and one by Transition US and both available online.[3] The model is based on the premise that climate change and peak oil are

[3] Based on the experience of Transition initiatives in the United Kingdom, Hopkins (2008:148–73) lays out "twelve steps" for creating a Transition initiative: setting up a steering group that would "design its own demise" when it had served its purpose; creating awareness through means such as film screenings and talks; laying the foundation for Transition by developing networks with all kinds of organizations in the community, including other environmental groups; organizing a "great unleashing" to kick off the Transition initiative; setting up smaller groups to work on different tasks and issues such as local food; using "open space technology" to hold idea-generating meetings; doing practical and visible projects such as planting fruit trees; holding "reskilling" events to teach people basic skills that are being lost, such as sewing; creating positive relationships with local governmental authorities; learning from older people through means such as oral history interviews; allowing the process to "go where it wants to go" and "seeing what emerges"; and creating an Energy Descent Action Plan.

urgent problems that will force drastic reductions in energy consumption. Communities need to start planning for an "energy descent" by creating resilient cultures and new structures. The model emphasizes the positive impact of local action in support of sustainable practices for individual empowerment and community relations.

The organizers of Transition Pittsburgh were entrepreneurial types who found the Transition model highly appealing. Greg Boulos had an organic farm outside Pittsburgh. Jeff Newman studied permaculture and organized a group called Sustainable Steel City. Mark Dixon, a film maker, learned about Transition while doing research for his environmental documentary, *Your Environmental Road Trip*. Elisa Beck, a fourth founder of Transition Pittsburgh, helped organize a group called Sustainable Monroeville in her community and later opened Schwartz's Market as a "living market" devoted to local food and sustainable living. Jeff attended a Transition training in New Jersey and met a national Transition trainer at a Pennsylvania Association for Sustainable Agriculture conference where she was giving a talk. He and others decided to invite her to Pittsburgh to do a two-day Training for Transition (T4T), which successfully attracted participants who were keenly interested in the Transition model. The organizers succeeded in applying for official status as a Transition Pittsburgh "initiative" of Transition US in February 2010.

On March 9, 2010, Transition Pittsburgh was launched with a well-attended showing of the *Transition 1.0* film (2009), after which attendees were invited to divide into groups and brainstorm about projects around topics such as food, energy, transport, economy, and creative land use. This successful movie screening and small group discussion was followed later by other film showings in a series organized by one of the founders. Transition Pittsburgh held another two-day T4T in June 2010, led by the national trainer who again came to Pittsburgh, which was attended by about twenty people eager to become involved. In an attempt to employ the "open space" model advocated by Transition (see Hopkins 2008:168–9), the group instituted weekly planning meetings at a café, dubbed Transition Tuesdays, held in the summer and fall of 2010. Several successful "gift circles" were also held, attracting twenty to thirty people each. The regular Tuesday meetings petered out, but in April 2011, one of the core Transition activists organized a "T4T Alumni Reunion and Open-Space Dinner" by sending an email to the list of people who had attended the Transition trainings. He then called another meeting for July 10, 2011 "to re-form the initiating group for the 3rd time" (email, June 28, 2011). This was the first of a number of twice monthly meetings held in Friendship Park during the summer, until they were moved to the back porch of a member who offered her nearby house as a more comfortable location. With the advent of Occupy Pittsburgh, which began in October 2011, she and some others became involved in Occupy and the Transition meetings were suspended. In December 2011, however, a young member of Transition, Justin, organized a lunch meeting at a restaurant to reconvene Transition "hub" members. Beginning

FIGURE 3.1 "Peaceful Gathering of Hands" in Pittsburgh's Schenley Park with Transition members and others (photo by Larry Rippel)

in January 2012, he organized twice monthly Transition potlucks, held at the First United Methodist Church, which also provided space for the Transition film screenings. The *Transition 2.0* film (2012) was first shown on May 21, 2012, followed by other films in the fall of 2012 at events that typically attracted more than thirty people.

Thus, Transition Pittsburgh went through a number of starts and stops, demonstrating both that it was difficult to stabilize the group and that many people continued to be interested in participating. At all of the gatherings, activists talked about a wide range of topics, including existing projects in Pittsburgh and ideas for new projects and strategies for implementing them. However, different people showed up from week to week, and there was little continuity or follow-through on ideas generated from one meeting to the next. The gatherings did provide networking opportunities, and some participants were able to connect with others who were interested in their projects. For example, a participant interested in aquaponics discussed the idea with others at a Transition event and eventually began a demonstration project. On the other hand, some participants were frustrated that a lot of ideas were generated at Transition gatherings, but little came of them; it typically took an entrepreneur to develop an idea into a project and not all ideas had entrepreneurs.

Transition's strategies reflected its loose organizational model and the reliance on entrepreneurs to generate actions. Transition members initiated a number of tactics ranging from major undertakings such as the "living market" and the aquaponics project to film showings intended to generate discussion and workshops designed to share skills such as sewing and bike repair. At Transition gift circles, participants exchanged material items, expertise, and services. One Transition member began an annual Peaceful Gathering of Hands held in Schenley Park that has attracted numerous participants (see Figure 3.1). As in the latter example, many activities were not official Transition events, but they were initiated by Transition members and loosely associated with the group. There were some attempts to define Transition projects more clearly and to create a more structured organization with standing committees and a steering committee. Most notably, members held a meeting in 2012 (described in Chapter 7) to clarify Transition goals, set up a structure, and define some projects. Despite this and other efforts, however, the suggested changes never took root. One problem was that persons who might have become key organizers did not have the time to devote to Transition. Another was that many members seemed to prefer the open space organizing style. Consequently, the group continued to rely on entrepreneurial actions and, eventually, the social networking tool Meetup to publicize events such as skill-sharing days.

Allegheny Group of the Sierra Club

The national Sierra Club is a long-standing environmental organization founded in California in 1892 as part of the early American conservation movement. The Sierra Club has numerous state chapters and regional groups throughout the country. The Sierra Club's Allegheny Group, one of ten regional groups comprising the Pennsylvania chapter, was founded in 1970 and currently has about 5,000 members (most of whom are paper members rather than active participants). In 2004, the national organization opened a field office in Pittsburgh with several staff members assigned to regional management and national campaigns. More recently, the state chapter also hired several staff members to run state campaigns who share the office. The local group held meetings and activities, such as weekly activist nights, at the field office.[4] As part of a larger national and state organizational structure, the AG also receives other resources, including a modest financial allotment based on membership, access to staff expertise, training opportunities, and national literature and campaign ideas. For example, the Pennsylvania state chapter not only hired a staff member to lead outings programs but also trained volunteers in Pittsburgh and other places around the state to help revive group outings programs (fieldnotes, June 11, 2018). At the same time as it receives support, the AG is also accountable to the state and national SC organizations

[4] In the fall of 2017, the Sierra Club field office moved to a shared office space in order, at the request of the national organization, to cut costs. The new office space could be used for meetings, but it was no longer a space belonging to the Sierra Club that could easily be used for a broader range of activities such as action nights.

FIGURE 3.2 Tabling by the Sierra Club on Earth Day 2016 (photo by Michael LaMark)

and must adhere to restrictions imposed by tax regulations and SC rules, which the group is careful to observe.

The AG is the most conventionally organized group in my sample. The group and chapter structures are dictated in large part by national Sierra Club requirements and models (e.g., bylaws), and each has an elected Executive Committee (ExCom) as its main decision-making body. In the AG, the ExCom meets once a month at a regular time, and a quorum of five out of the eight-member ExCom (chair, vice chair, secretary and treasurer plus four others) is needed to make decisions. The group uses parliamentary procedure for all official decisions, such as approval of the minutes, and puts off decisions, small or large, when there is no quorum of ExCom members at a meeting. In addition to the ExCom, the group follows the national SC model in having a number of issue chairs (air quality, coal power plants, conservation, mining, political, transportation, and water quality), but added a local innovation of several action teams (endangered species, global warming, public lands, and shale oil and gas) in 2009. Meetings of the ExCom are open to all members of the Sierra Club, and during my observations, they were typically attended by ten to fifteen people, including some issue or action chairs and former ExCom members who remained active in the group, often as issue chairs, creating some continuity in leadership and potential for learning and assessment of past strategies and tactics. The staff member assigned to the national Beyond Coal campaign among other responsibilities, Randy Francisco, regularly attended the ExCom meetings during the time of my observation and tried to assist the volunteers with advice and information. Beginning in 2017, a new state SC staff member, Eva Resnick-Day, began attending ExCom meetings regularly. For some time,

FIGURE 3.3 Sierra Club participation in the 2011 St. Patrick's Day parade (photo by Maren Cooke)

Randy also ran weekly "action nights" at which volunteers made phone calls and did other tasks, sometimes to assist him with his work and sometimes to encourage members to participate in activities initiated by volunteers. Later, volunteers took over running of the action nights at their request, hoping to expand the group by working on a wider range of issues.

This structure has maintained the Allegheny Group for many years as a stable organization. The local group has been active over the years on a range of issues such as preservation of public lands, coal mining, shale gas drilling, global warming, energy conservation, and endangered species. Members have regularly engaged in activities such as tabling at Earth Day celebrations (see Figure 3.2) and participating in events such as St. Patrick's Day parades (see Figure 3.3). Some of the group's activities are part of state and national

SC campaigns, such as the effort to regulate mercury emissions from coal-fired plants, and in some instances, staff have helped to organize volunteers. In 2006, the AG worked on a successful campaign to get the Pennsylvania Department of Environmental Protection (DEP) to make mercury pollution from coal-fired plants and other sources a priority. The AG regularly works with other organizations in the region, such as the Center for Coalfield Justice, Clean Water Action, PennEnvironment, PennFuture, and the Group against Smog and Pollution (GASP), and often activates its members to testify at hearings and show up at events. The AG's political committee interviews political candidates and solicits information on a questionnaire so that the group can endorse candidates and work on elections regularly. Before endorsing a candidate within Pennsylvania, the group must get the approval of the state chapter, while national-level candidates are approved by the national organization. The group also supports the choices of the national SC; for example, chapter and group members worked for the election and reelection of President Obama in 2008 and 2012 and for Hillary Clinton in 2016.

Marcellus Protest

The group that came to be known as "Marcellus Protest" originated when neighbors began talking to one another and organizing in their communities as they learned about the leasing of land for Marcellus Shale drilling in and around Pittsburgh. As described in Chapter 4, residents were getting offers to sell their drilling rights in areas of the city, such as Lawrenceville and Lincoln Place. After investigating the issue of fracking and becoming alarmed about the possibility of it happening in their communities, Lincoln Place residents Mark and Elizabeth Schneider began holding meetings in their home of what became the Lincoln Place Action Group (LPAG). The group then organized a town meeting at Mifflin School in Lincoln Place, which attracted a large crowd. Ken and Loretta Weir, residents of Lincoln Place who attended the meeting at Mifflin School, later organized a meeting at the Elks Club in West Mifflin on July 27, 2010 that was attended by some of the activists who founded Marcellus Protest.

After these early meetings, an ad hoc group of experienced activists, newly mobilized neighborhood activists, and other concerned citizens decided to try to build the movement against fracking by organizing a large demonstration in Pittsburgh on the occasion of a shale gas industry conference at the convention center in downtown Pittsburgh. Originally, the industry conference was scheduled for October 1, 2010, but was later moved to November 3, 2010, giving activists longer to organize. On September 3, 2010, organizers held the first of a series of public meetings to plan the protest, which were each attended by about twenty to fifty people. A number of working groups were formed, such as Outreach, Media, Logistics, Fundraising, and Art.

The working groups were extremely active, attending to all of the details needed for a large demonstration and allowing individuals to contribute their skills. The Art group, led by two talented artists who enlisted many others to help them, created large, colorful banners, puppets, and signs. The Fundraising committee, led by Dana Dolney, held a number of events that allowed for participation and fun, while raising money. These included a very successful "Shale Trail" awareness night and fundraiser at a bar crawl in the Lawrenceville area of Pittsburgh. The Outreach committee worked on making contacts with other groups, such as environmental organizations, and sent out letters asking for endorsements and contributions for the November 3 march. Organizers also worked with the Pittsburgh Student Environmental Coalition (PSEC) to organize students and show *Gasland*, a critical organizing tool, at area colleges. Members of the Shadbush collective were very active in organizing door-to-door outreach, using their "What the Frack?" pamphlet, which explained the process of hydraulic fracturing and detailed its potential impacts on human health and the environment.

Organizers were rewarded for all of their work with a memorable November 3, 2010 demonstration that was an upbeat event on a beautiful fall day. At the start of the march, speakers and performers rallied demonstrators who were gathering at the Allegheny Landing on the North Shore of the Allegheny River. Hundreds of activists, accompanied by a marching band, then crossed the Rachel Carson Bridge (named in honor of the Pittsburgh-area environmental pioneer) to the David L. Lawrence Convention Center in downtown Pittsburgh, where the gas industry conference was in progress (see Figures 3.4 and 3.5). At this site, more performers and speakers inspired the protesters, including *Gasland* producer Josh Fox and city councilor Doug Shields. Participants from all over Pennsylvania and New York, as well as other states, joined local Pittsburgh activists in the emerging movement as they spent the afternoon talking, distributing information, selling t-shirts and buttons, and building ties with one another.

The demonstration was considered highly successful, and a few weeks later, on November 16, 2010, the Pittsburgh City Council voted unanimously to become the first US city to ban shale gas drilling. Anti-fracking organizers had been lobbying for the ban for months, attending city council sessions on the issue in the summer as well as holding community meetings. The ban's sponsor, Doug Shields, believed the grassroots pressure was critical to passage of the city ban (interview, April 3, 2012). As a result of the demonstration and the ban, the Pittsburgh movement gained a great deal of recognition in Pennsylvania. Marcellus Protest also became known elsewhere in the United States and, to some extent, internationally owing to the excellent website created by David Meieran, one of the original organizers of MP, and to Pittsburgh's status as the first city in the nation, perhaps in the world, to ban shale gas drilling based on a community rights ordinance.

FIGURE 3.4 "We Can't Drink Money" banner at head of the Marcellus Protest march on November 3, 2010 (photo by Tom Jefferson)

FIGURE 3.5 Marcellus Protest march across the Rachel Carson Bridge on November 3, 2010 (photo by Tom Jefferson)

Organizers immediately worked to build on the inspiring success of the demonstration and the victory of the Pittsburgh ban on fracking. To follow up on the November 3, 2010 event, MP held an open meeting on November 12 to ask, "Where do we go from here?" The organizational structures used to mobilize the demonstration provided a model for an ongoing organization, and participants in the meeting managed to create a basic structure that included a coordinating committee, general assemblies, and working groups. However, it was difficult to work out details about how the group would function at this large meeting at which participants with different backgrounds and experiences and few common ideas about how to organize tried to plot a future direction for both an organization and a movement (fieldnotes, November 12, 2010). From 2010 to 2014, MP held a series of general assemblies as the group tried to expand the movement. As detailed in Chapter 7, some of these meetings attracted newcomers and produced promising strategic ideas while others were less successful.

After the big November 3 march, activists continued to organize smaller-scale demonstrations, typically when individuals took the initiative to organize them. On November 4, the next day, Loretta and Ken Weir organized a demonstration at the riverside to protest the illegal dumping of fracking water. In early December 2010, activists protested at the Carnegie Museum in Pittsburgh in response to the possibility that the museum would lease land for fracking that it owned outside the city and used for scientific research, and apparently in response to the protests, the museum soon decided not to do so. A few days later, activists led by Loretta Weir sang anti-fracking Christmas carols in downtown Pittsburgh on the steps of the City County Building. In January 2011, Gloria Forouzan put a great deal of energy into organizing a bus to go to Harrisburg to join other activists from around the state to protest the inauguration of pro-fracking Governor Tom Corbett. Unfortunately, icy weather prevented many from attending, leaving the bus only partly filled, and for those of us who traveled to Harrisburg, it was a very uncomfortable demonstration in the freezing cold. Nevertheless, connections were made with activists across the state, and the Pittsburgh participants were applauded at a large meeting following the demonstration (fieldnotes, January 18, 2011). Pittsburgh was becoming known as a movement center after staging the large demonstration, launching a highly informative and frequented website, and passing the first US city ban on fracking. Activists from Pittsburgh interacted with those from other regions as they participated in protests and lent their expertise to the movement and as activists from other regions visited Pittsburgh.

In 2011, Pittsburgh residents attended more rallies in Harrisburg and testified at hearings around the state. Marcellus Protest worked with others in the movement community, raising the issue of fracking along with the Sierra Club, for example, at the annual St. Patrick's Day parade in Pittsburgh in 2011, and causing a stir among parade organizers who withdrew the Sierra Club's invitation to the parade the following year. The Thomas Merton Center, a local peace and justice center, gave its annual "New People Award" to Marcellus Protest in 2011. Activists worked to ban fracking in other municipalities, with some successes, including neighboring Wilkinsburg. When Occupy Pittsburgh formed in October 2011, Marcellus Protest members carried their banner in the inaugural march (and a few began to spend more time with Occupy than with MP). Marcellus Protest joined Shadbush activists and participants from the Occupy camp for a demonstration at a second annual gas industry conference in November 2011. In December, there was more Christmas caroling at the City County Building.

In 2012 and 2013, MP continued to participate in demonstrations, hearings, and supportive actions in other parts of the state as well as in Pittsburgh. Often, MP worked with Shadbush to organize actions and events. On March 20, 2012, Marcellus Protest and Shadbush held a demonstration and educational forum focused on pipelines in response to a gas industry conference on Marcellus Shale infrastructure. In July 2012, the two groups put together a demonstration for the Pittsburgh leg of the "Tour de Frack" bicycle ride to Washington, DC organized by Marcellus Outreach Butler (MOB), the Butler County group that came out of Marcellus Protest. On Earth Day 2013, Marcellus Protest and Shadbush worked together on a demonstration at the office of the DEP, along with a statewide coalition that organized demonstrations against the DEP across the state. The DEP was targeted for not adequately protecting communities against fracking, but it was also an attractive geographic target in Pittsburgh. Because the DEP office in Pittsburgh is located on an island, Shadbush activists organized an "Allegheny Armada for Clean Water" of kayaks and canoes (despite the freezing water) to meet up with about 100 other demonstrators who marched across a bridge for a brief occupation of the DEP (see Figures 3.6 and 3.7). In presenting their demands to the DEP, protesters sent the message that citizens were concerned about fracking and wanted protection from their government.

Following this memorable demonstration, Marcellus Protest joined with a number of groups on an organizing committee to bring Josh Fox and *Gasland II* to Pittsburgh in June 2013. As a result, some 1,700 persons attended the showing in the city, the final stop on Fox's nationwide tour before the debut of the documentary on HBO. Beginning in July 2013, MP became heavily involved in the Protect Our Parks campaign against fracking under Allegheny County parks, which helped revitalize the local movement. In October 2013, MP and Shadbush demonstrated in support of POP with Power Shift, a youth climate summit, when it came to Pittsburgh. In 2014, a number of activists from MP were involved in creating a new statewide coalition, Pennsylvanians Against Fracking, organized with the initial goal of winning a wider moratorium on fracking. Marcellus Protest members also became involved in other groups such as Friends of the Harmed, which produced two volumes of

FIGURE 3.6 By land and by sea: the Allegheny Armada for Clean Water (photo by Tom Jefferson)

FIGURE 3.7 The march over the footbridge to protest the Department of Environmental Protection on Earth Day 2013 (photo by Tom Jefferson)

Shalefield Stories, compilations of the testimonies of people who have been harmed by shale gas drilling (www.shalefieldstories.org).

Although a number of the original MP activists went on to work in statewide groups and to help people adversely affected by fracking, they remained part of the network and often showed up for tactics such as testimony at county council meetings, even if they didn't attend the meetings of groups such as MP or POP. Marcellus Protest stopped holding meetings in 2013 but remained in existence as a network, and some individuals continued to receive invitations to represent MP as speakers or participants in events. Marcellus Protest Google groups, which David Meieran set up in 2010 to allow for email exchanges, continued to be used to share information and calls for action in the network. The MP website, Facebook page, newsletter, and calendar, which covered events across the state, were also important tools for the movement. A few activists, particularly Gloria Forouzan and John Detwiler, maintained these, so there was at least a continuous online presence that allowed new people to find the movement. In fact, people from around the world contacted MP through the website, and these contacts helped connect Pittsburgh activists to others, both nationally and internationally.

Shadbush Environmental Justice Collective

The Shadbush Environmental Justice Collective was formed in 2010 after some young activists in Pittsburgh became involved in protests against mountaintop removal. This happened after a group of Mountain Justice activists came through the city and participated in a roundtable discussion with local activists. The event had a big impact on local participants, who decided there was a need for a radical environmental group in Pittsburgh. At the same time, activists were becoming involved in the shale gas drilling issue in a small group that merged with Shadbush. Thus, compelling environmental issues, including threats from fossil fuel industries and climate change, inspired the formation of a radical environmental group. Although Shadbush took a multi-issue approach, most of its activities centered around fracking, which became a "hot" environmental issue in the region soon after Shadbush's founding. In this way, the group was similar to Marcellus Protest, and several members were active in MP, but Shadbush was a more ideologically cohesive group.[5]

The Shadbush collective valued participatory democracy and strove to create an organization that reflected the ideology of its members. Participants were all

[5] In the *Shalefields Grassroots Reporter* (Summer 2011, p. 7), the group explained its radical perspective with this statement: "The Shadbush Environmental Justice Collective works to confront the root causes of ecological destruction, climate change and oppression. We are committed to building community-based, radically sustainable solutions that foster local autonomy and self-sufficiency in Pittsburgh and its bioregion."

young people in their twenties and thirties (except for me, who they kindly allowed to join the group as a participant observer), and all were white.[6] The group had to deal with the problem of how to retain the tight-knit relationships and ideological coherence of the group while also welcoming new members. Shadbush was small, with about six to eight active members at a given time, and it was exclusive insofar as potential members had to agree with a statement of principles and were typically known to one or more members of the collective. This reduced the likelihood of internal conflict, but also kept the group small and limited its strategic capacity. In addition, the collective met frequently and required a high level of commitment. Members recognized that not everyone wanted to make this type of commitment and, as detailed in Chapter 5, they discussed ways to create another tier of membership among supporters who would like to be involved and help with projects without becoming core members of the collective. To some extent, this happened informally because group members had extensive networks of friends and radical activists, many of whom turned out for various Shadbush events and contributed particular skills when asked to do so. However, the only formal means of connecting supporters implemented was a larger email list of Shadbush supporters.

Owing in part to its extensive networks and also to the energy of core members, Shadbush was quite active, resulting in a fair amount of visibility and growing credibility among radical environmentalists in Pennsylvania and beyond. The group was an affiliate of the radical environmental network Rising Tide North America, and a few Shadbush members traveled to some Rising Tide gatherings and regularly participated in conference calls. Shadbush also participated in MP, though this participation lessened as MP struggled to maintain itself and Shadbush became more active in its own right. The collective developed some important educational materials, including a "Fracking 101" presentation intended to explain the fracking process and its consequences to audiences unfamiliar with the issue. The Shadbush collective also put out two issues of a newsprint publication called the *Shalefields Grassroots Reporter* in Summer 2011 and Spring 2013, with financial support from MP. The first edition laid out the problems, while the second one pointed to actions being taken, and both were widely distributed and very well received. Production of the *Reporter* took an enormous amount of time and effort, and both issues ran way behind schedule. Although the collective was proud of the end

[6] As an environmental justice group, Shadbush was concerned about class and race issues in connection with environmental hazards, but did not have much capacity for addressing them. Shadbush talked from time to time about how the collective might diversify, recognizing the limitations of their small, all-white group. One Shadbush member was in touch with a leader of New Voices for Reproductive Justice, a women of color group focused on health and environmental concerns, and in 2014, Shadbush split the proceeds of a fundraiser with New Voices and Water for Woodlands, a support group for Butler County residents who had lost their water supply to fracking.

results, at least one Shadbush member questioned the advisability of such a time-consuming project.

Beyond these educational efforts, which reflected Shadbush's ideological position, the group worked to spread direct action to the movement.[7] In October 2011, Shadbush presented a direct-action workshop, open to allies in the anti-fracking movement and intended to spark direct-action tactics. However, it was only attended by about six persons other than the Shadbush members who organized it and did not result in the planning for a direct action that the group had envisioned. Soon afterward, the collective worked with MP and recruits from the Occupy Pittsburgh camp to stage a demonstration at a "developing unconventional gas" industry convention, known as DUG East, in downtown Pittsburgh. The action attracted quite a few participants, owing in part to the presence of the Occupy Pittsburgh camp, and resulted in a handful of arrests. Although this was considered by many a successful action, the collective felt that they had not been able to control the planning of it adequately and that reliance on people they didn't know well created problems. As was typical of Shadbush, the group "debriefed" on the action afterward and changed its future strategy as a result of lessons learned.

Shadbush continued to practice direct action. In May 2012, the group worked with Marcellus Outreach Butler to stage an "action" wearing hazmat suits at the office of Brian Ellis, a state legislator who sponsored Act 13, a law that favored gas companies over local communities. MOB consisted largely of older retirees, and the Shadbush activists were young people who knew how to engage in direct action; working together, they formed an effective team, and enduring new relationships were created. In addition to participating in MOB's Tour de Frack bike ride in the summer of 2012, Shadbush organized several tours of fracking sites in heavily drilled areas and developed relationships with some of the people affected by fracking in Butler County who had lost use of the water in their homes (see https://sites.google.com/site/waterforwoodlands). Shadbush also put on a hoedown fundraising dance that raised $2,500 for the Butler County families in October 2012 (followed by another similar fundraiser in April 2014). In November 2012, Shadbush held an "action camp" at an organic farm owned by the Henry family. Distressed by the construction of a nearby shale gas well, the owners invited Shadbush to hold a weekend camp at their farm, dubbed "Maggie's Farm" for "the farmer's wife" (who sold eggs under this name), Maggie Henry. The action camp, which included workshops and panels of speakers from Pennsylvania, Ohio, and New York, was attended by many people from anti-fracking groups such as MOB and Marcellus Protest. Like many such camps, it ended with an "action," which was a rally held at the drilling site in Lawrence County.

7 Direct action consists of tactics such as demonstrations to pressure targets directly rather than employing institutionalized tactics such as electoral politics or lobbying.

FIGURE 3.8 Shadbush protest at Shell fracking site with giant pig in January 2013 (photo by Tom Jefferson)

The action camp involved a huge amount of work, but the collective and their friends did a great job on the logistics, attracted a large number of attendees, and further enhanced Shadbush's reputation. Again, the group analyzed the positives and negatives of their work afterward, expressing disappointment with the action because they felt that they just stood around with signs and did not accomplish much. But unlike action camps held by other radical groups, which typically culminate in a one-shot action, Shadbush decided to return to the drilling site near the farm for another, better planned and more extensive action in late January of 2013. For this action, the group mobilized a larger circle of activists, including some who had attended the action camp and some from the Butler group (many of whom also attended the action camp). Maggie Henry also participated in the Shadbush-led protest against Shell alongside the other activists (Crompton 2013). The protest involved months of planning and training for roles such as media spokesperson and police liaison. It also included the creation of a large paper mache pig called "Henrietta" (representing the Henry family's organic farm, which raised pigs), which surprised the police when it was unloaded from a truck. Several activists locked themselves to the pig and blocked the road to the fracking site (see Figure 3.8). The carefully planned action resulted in much media publicity and no arrests, owing to negotiations with the police.

The "pig" action created even more solidarity within the group and with its allies and added to Shadbush's growing reputation as an effective and highly active group. When the statewide coalition of groups decided to protest at the regional offices of the Pennsylvania DEP on Earth Day 2013, local activists were eager to have Shadbush join in the planning. Attracted by the idea of a brigade of kayaks and canoes, Shadbush decided to volunteer for a lead role in organizing the protest. Like the action with the giant pig, the Earth Day demonstration was considered a great success that fired up protesters and garnered much media attention. Working with other groups and a network of activists, Shadbush was succeeding in organizing a series of protests that helped mobilize the larger movement.

Shadbush talked about strategy frequently, holding numerous meetings and other strategy sessions. The group's identity as a radical environmental justice group predisposed it to certain types of tactics, mostly but not exclusively direct-action tactics. Some Shadbush members had experience in other groups such as Mountain Justice, Earth First! and Occupy Pittsburgh, and they frequently brought these experiences into strategic discussions in the collective. One explicit goal adopted by the group in 2012 was to introduce more direct action to the larger movement, which was accomplished in collaborations such as the one with the Butler group. Another explicit goal, adopted about a year later, was to have an ongoing campaign rather than isolated actions. In the spring of 2013, Shadbush hosted a campaign planning workshop and several meetings, inviting allies from MP and other groups to plan a multiaction campaign aimed at selected targets. That effort resulted in several campaign ideas, but one new development proved irresistible as the focus of a new campaign. When Allegheny County executive Rich Fitzgerald announced his intention to lease the county parks for fracking, activists quickly agreed that this should be their focus. A new coalition, called Protect Our Parks (POP), launched a campaign targeted at the county executive, which mobilized new participants and reinvigorated the local anti-fracking movement.

Particularly when its direct-action skills were useful, Shadbush participated in POP, as in the case of the demonstration and sit-in with Power Shift in October 2013 and a demonstration at Deer Lakes Park in April 2014. However, the collective ended up taking a supportive rather than leading role in the campaign, which mainly used tactics of lobbying county councilors and testifying before the county council. While Shadbush did not object to these tactics, they were not part of the group's usual repertoire, and most members found themselves uncertain about how to contribute to the campaign. Shadbush continued to hold irregular meetings in 2014 and to engage in other activities, including a third retreat in February 2014 and a second hoedown fundraiser for the Woodlands in April 2014. By the end of the year, however, the group was petering out as core members left town or had to decrease participation because of their jobs, and the group disbanded to give rise to Three Rivers Rising Tide.

FIGURE 3.9 Protect Our Parks demonstration with Power Shift, October 19, 2013 (Copyright ©, Pittsburgh Post-Gazette, 2019, all rights reserved. Reprinted with permission)

Protect Our Parks

Protect Our Parks was a coalition of organizations and individuals that began in August 2013 in response to the announced intention of the Allegheny County executive, Rich Fitzgerald, to lease county parks for fracking (McGill 2013). The fracking was to occur under rather than in the parks, he said, and Deer Lakes Park, located in a relatively out of the way, nonaffluent part of the county, would be the first to be leased. Earlier in 2013, the county had leased land around Pittsburgh International Airport after a brief public debate demanded by anti-fracking activists. A couple of county councilors who voted in favor of the airport lease later told protesters that they regretted their votes, leading activists to think that if they could only change the minds of a few councilors, they could win a victory on the parks. Moreover, they expected leasing of public parks to arouse a lot more public outrage than leasing land around the airport.

The POP coalition got its start after Shadbush members discussed at their February 2013 retreat the desirability of spearheading a campaign rather than continually engaging in one-off actions. Recognizing that they had limited capacity to organize a campaign on their own, the Shadbush collective decided to invite a variety of activists from other environmental groups, including MP and regional anti-fracking groups, Mountain Watershed Association, Clean Water Action, and the Sierra Club, to a brainstorming session to come up with ideas for a campaign. They wanted the campaign to bring together different groups and individuals, helping to build the movement and, hopefully, produce a victory. At an initial meeting, a number of ideas were generated, and participants agreed to meet again to discuss them further; in between the two meetings, the county executive declared his intention to lease Deer Lakes Park for fracking, and the group decided to create a campaign around this immediate threat. The next meeting was a planning meeting for the new coalition, followed by numerous other meetings, which attracted a variety of participants. Early meetings were attended by fifteen to twenty people, including some who had not come out for MP meetings in some time but who were part of the MP network. Later, the core of meeting participants shrank to about ten regulars, but many more people participated in campaign tactics and events between 2013 and 2016.

In August 2013, the campaign was publicly launched when county residents began attending meetings of the Allegheny County Council and speaking out against fracking under the parks.[8] POP drafted an open letter to county councilors opposing the leasing of public parks and collected endorsements of organizations for the letter. The group also distributed a sample letter for individuals to use as a model in writing to their own county councilors. A number of POP participants set up meetings with individual county councilors to lobby them, and they encouraged residents to call and write to council members. Many people spoke at the county council meetings, even before the county issued a Request for Proposals (RFP) from companies to drill and later awarded a contract to Range Resources, the only bidder. POP developed a flyer about the issue, which group members distributed widely at various events and venues in the city. The group also created a petition against fracking in the parks, which POP members used to explain the issue to citizens as they collected signatures while tabling and flyering at various public events. Many people continued to testify before the county council, often late in the evening. POP also provided county council members with extensive educational materials about the health and environmental risks of fracking. POP members were developing relationships with county councilors, including allies, opponents, and some open to persuasion. One important ally, Barbara Daly Danko,[9] sponsored legislation in September 2013 calling for a three-year moratorium on drilling in or under the parks. POP supported this initiative as a first step, but the measure failed to gain the support of the council.

While POP focused much of its energy on lobbying, petitioning, and testifying, the coalition also engaged in some direct action, usually led by Shadbush. The demonstration with Power Shift consisted of an unpermitted march with hundreds of participants, which began with a skit using a large puppet of Rich Fitzgerald outside the convention center and then proceeded to the City/County Courthouse for speakers and a rally. A small group staged a 3-hour sit-in at the

[8] This was done even before the matter was before the county council by signing up for 3-minute time slots at the end of the meeting during the period reserved for public comment on items not on the meeting agenda.

[9] Sadly, Barbara Daly Danko died of cancer at age 61 on May 6, 2015, and the movement lost its strongest ally on County Council.

office of Rich Fitzgerald, which attracted media attention and resulted in no arrests before participants finally declared victory and left. POP was grateful to Shadbush for organizing the Power Shift demonstration, and Shadbush was happy to have contributed in a way that utilized the group's skills given that lobbying activities were not very attractive to its members.

POP worked hard to build support and to inform the public about fracking of the parks. At the end of October 2013, POP held a town meeting with State Senator Jim Ferlo at a school in the area near Deer Lakes Park. That effort attracted some residents of the Deer Lakes Park area, and in early January, POP held a public meeting to include newcomers in the campaign that attracted some of these residents as well as other new faces. In late January 2014, POP joined the Sierra Club for phone banking to try to get Club members to call their county councilors. POP also explored various strategic options, including the possibility of holding a countywide referendum against fracking in the parks. Barbara Daly Danko attended a POP meeting to argue that POP should consider a referendum, but the group decided that they lacked the capacity for this action, which would involve collecting thousands of signatures of registered voters. Moreover, they feared that if a referendum was defeated, it would be a major setback for the movement. In December 2013, POP was buoyed by a Pennsylvania Supreme Court ruling on Act 13, legislation on natural gas operations. In response to a challenge by environmentalists, the Court called for "due diligence" by government officials in investigating the impacts of fracking before permitting it. This directive "became the cornerstone of POP's message to Allegheny County Council" as the group continued its testimony (Detwiler 2014:5). By December 2013, POP had spent so much time testifying before the county council that the group brought a cake to the final county council meeting before the holiday break in a gesture of goodwill to thank the councilors for listening.

In 2014, as POP waited for the introduction of legislation to allow fracking under Deer Lakes Park, the group called for public hearings and continued to testify at council meetings, wearing bright green scarves made by a POP member to make the activists visible in the county council chambers. POP also held fundraisers and distributed informational materials in numerous venues, accumulated organizational sponsors, wrote letters to the editors, put out press advisories, contacted reporters, and held press conferences. A number of POP members were in regular contact with county councilors, building personal relationships with them. In April 2014, the county executive finally submitted legislation to the county council to authorize a lease for drilling under Deer Lakes Park. On April 2, Rich Fitzgerald held a "public meeting" on the proposed fracking at Deer Lakes High School, which was attended by about 500 people, including both POP supporters and allies of Fitzgerald. POP members spoke at this meeting as well as at the next county council meeting and a hearing held by the county council on April 15. On April 4, Pittsburgh Artists Against Fracking held a benefit for POP. On April 19, POP held a

"Hands Around the Lake" event at Deer Lakes Park attended by about seventy supporters who paraded around a lake in the park, distributed flyers, and enjoyed the outing on a beautiful spring day. In short, POP was very active, with participation from member groups such as the Sierra Club and from highly committed individuals, and many events helped build solidarity among activists.

At the May 2014 meeting at which the county council voted on the legislation, POP presented its written "Case against Fracking Deer Lakes Park" and its petition with more than 7,000 signatures, the pages of which were taped together and used by activists in an impressive presentation circling the room. Despite testimony by a large number of Allegheny County residents opposed to the measure, the council voted in favor of leasing the park by 9 to 5 (with one abstention). The scene was dramatic, with many POP supporters speaking passionately and one council member objecting to the accusation that the county council had not done "due diligence" by petulantly stomping out of the room at one point. Later, Loretta Weir thrilled POP members by scolding him memorably during her three minutes, telling him it was his job to listen to citizens. It was quite late at night before the vote was taken, and POP members waited along with members of the press in the packed chamber. In the end, the female and black members of the county council voted against the legislation (or abstained), while the white men on the council all supported the county executive. Some POP members yelled "shame," and a designated spokeswoman read a statement from POP for the media (fieldnotes, May 6, 2014).

The defeat was heartbreaking for POP, which had put so much time and effort into trying to prevent the infringement of fracking into public parks and hoped that the campaign would produce a victory for the movement. The group then had to figure out what to do next. A lawsuit on the grounds that the council had not exercised due diligence was considered carefully, but the idea was finally abandoned after a nonprofit environmental law firm advised the group that the case would be unlikely to succeed and would be prohibitively expensive. Instead, POP decided to collect signatures that would allow the group to sponsor a citizens' ordinance calling for a two-year moratorium on further leasing of county parks for fracking. POP framed it as a "wait and see" ordinance, a prudent step until the impact of fracking on Deer Lakes Park could be assessed. The tactic involved a great deal of research into the legal requirements for a citizens' ordinance as it was the first time the provision for citizen-sponsored ordinances in the Allegheny County Home Rule Charter had ever been used. It also required a big effort to collect signatures. Although only 500 valid signatures were required, POP wanted to be sure to have plenty of extras in case any were challenged. More than forty-five volunteers working with POP ended up getting about 1,800 signatures, and the county did not challenge their validity. The tactic also involved a great deal of work in actually drafting the ordinance in consultation with helpful county staff members and with council ally Barbara Daly Danko. Two of the activists who worked extensively on it, Joni Rabinowitz and John Detwiler, wrote an op-ed framing the ordinance as a common-sense measure, which was published in the *Pittsburgh Post-Gazette* after the newspaper had editorialized against the ordinance (Rabinowitz and Detwiler 2014). POP members again met individually with county council members, but in the end, the council voted down the ordinance on the grounds that it would tie their hands unnecessarily.

POP was pursuing other tactics simultaneously with the citizens' ordinance and did not give up after this second major defeat. The group began to target the Allegheny County Health Department (ACHD) in the spring of 2014 on the grounds that the health department should be addressing fracking along with air pollution from sources such as Shenango Coke Works. Ted Popovich, a member of POP who was also a member of GASP, was heavily involved in this issue, and he regularly attended ACHD meetings. In September 2014, an opportunity to target the ACHD arose when the health department began holding public meetings in each of the thirteen county council districts about health issues important to the county as part of the ACHD's effort to achieve national accreditation from the Public Health Accreditation Board (Riely 2014). POP members began attending the meetings and sent a letter to the ACHD director, urging her to pay attention to the health impact of fracking. Later, the group sent letters to ACHD board members, held a press conference, and sent a letter to the accreditation board about the failure of the ACHD to address citizen concerns.

In addition to continuing to target the ACHD, POP kept an eye on developments at the Deer Lakes Park drilling site and declared itself a watch dog group. On May 30, 2015, POP held a "Protect Deer Lakes Park Day" at the park to monitor developments and get together for a picnic. Some members brought their dogs, wearing "watch dog" signs, and informational flyers were distributed, asking park visitors to watch out for problems (see Figure 3.10). In the summer of 2015, POP decided to have some fun with Rich Fitzgerald, who was running virtually unopposed for reelection in the fall. POP started a write-in campaign to elect a dog named Nellie (the dog of a POP member) for county executive, developing literature to explain why Nellie would be a better county executive than Fitzgerald. This campaign, which gained a following on Nellie's Facebook page, continued into the fall as POP supporters enjoyed promoting the campaign before writing in Nellie's name on their ballots. While having fun with that campaign, POP also tried to determine which other county parks might be targeted next and to make contact with community groups in the areas around county parks. With the slowdown of the gas industry, however, no further plans for leasing were announced by the county executive. While the drilling site at Deer Lakes Park was prepared, work was halted and no actual "fracking" had been done by the summer of 2016. POP continued to meet up until July 2016 and then decided to suspend meetings until threats to the parks

FIGURE 3.10 Protect Our Parks members gather at Deer Lakes Park as "watch dogs," May 30, 2015 (photo by Dianne L. Peterson)

resumed. Fracking under Deer Lakes Park did occur later in 2016, but no other county parks were leased, and POP did not begin meeting again.

Although POP lost the battle for Deer Lakes Park and ceased meeting, the campaign did strengthen the local movement community. Core members of POP formed a caring group of people with a genuine sense of community, which was on display when POP responded to the tragic loss of one of its own in June 2015. George Jucha, Jr. was a core member of POP who was also becoming active in the Sierra Club and Pittsburgh 350.org before he was killed suddenly in an automobile accident at the age of 50. George attended meetings regularly, testified before the county council, and volunteered his considerable skills as a writer and photographer to produce literature for POP, among his many other contributions. About a week before his death, he attended the picnic at Deer Lakes Park as did one of the county councilors who had voted against leasing the park for drilling, Heather Heidelbaugh. Many POP members and other local environmentalists attended George's funeral and expressed condolences to his parents, who asked that contributions in George's honor be made to POP. Protect Our Parks members joined with Heather Heidelbaugh to organize an event at the county council at which Heidelbaugh introduced a proclamation for "George R. Jucha, Jr. Day" using a script written by a POP member. POP members, along with George's family, attended the meeting wearing their trademark green scarves. In addition to this gesture, members of POP kept in close contact with George's parents, who later attended a meeting of POP followed by a lunch with the group.

TABLE 3.1 *Sample organizations*

Organization	Dates active	Goals	Organizational structure	Types of participants	Strategies and tactics
Transition Pittsburgh	2010–13 (continued using Meetup through 2016)	Reduced consumption and transition to fossil-fuel-free economy to address peak oil and climate change; create local models of sustainable living and food production	Network of local projects; open space meeting; loose connections to national and international organization	Green entrepreneurs; persons engaged in projects or seeking to become involved in creating models of sustainable living	Education through films and speakers; local sustainable initiatives; gift circles, skill-sharing events
Allegheny Group of the Sierra Club	1970–present	Elect environmentalists to public office; address wide range of environmental concerns, including air and water pollution, preservation of public lands, impacts of coal mining and fracking, endangered species, and climate change	Formalized structure with Executive Committee and issue committees; formal connections to state chapter and national organization	5,000 paper members; active core of about thirty environmentalists	Electoral politics; tabling, petitioning, lobbying, testifying, demonstrations, and coalition work; volunteer work on recycling and community gardens

(continued)

TABLE 3.1 (*continued*)

Organization	Dates active	Goals	Organizational structure	Types of participants	Strategies and tactics
Marcellus Protest	2010–13 (continued to exist as a network; meetings held in 2019)	Stop fracking in the Marcellus Shale region; ban fracking in Pittsburgh and other municipalities	General assemblies, working groups, and coordinating committee	Residents of threatened areas and other concerned citizens and environmentalists; both experienced and new activists	Demonstrations, direct actions; public meetings and educational work
Shadbush Environmental Justice Collective	2010–14	Address root causes of ecological destruction, climate change, and oppression; multi-issue but focused on banning fracking	Collective with consensus decision-making; loosely affiliated with Rising Tide North America	Radical environmentalists, mostly young with movement experience	Organizing and educational work, production of materials such as broadsheets; fundraisers; direct actions
Protect Our Parks	2013–16	Ban fracking in or under county parks	Coalition with initial meetings facilitated by Shadbush and others; later collective-style core group meetings	Environmentalists representing multiple groups in Pittsburgh	Letters to county council and media, meetings with county councilors, testimony at county council meetings; tabling; petitions; citizens' ordinance; demonstrations

Summary

The sample organizations described in this chapter represent a range of different types of grassroots environmental groups. The types of participants in these groups and their activities reveal that there are many different forms of grassroots activism in the environmental movement. Table 3.1 summarizes the characteristics of these groups. Transition Pittsburgh, which met irregularly from 2010 to 2013, is an outlier insofar as its purpose was to change the culture by creating models of sustainable communities rather than engaging in directly political action. The group aimed to be welcoming of participants with a range of political ideologies, and many of the participants were entrepreneurial types as well as persons interested in alternative lifestyles and values. Transition Pittsburgh had a very loose structure with a changing cast of leaders, and it held "open space" meetings. Its strategies and tactics attempted to follow the international Transition model of creating educational forums such as film series and engaging in activities such as skill sharing and practical projects. The Allegheny Group of the Sierra Club, which was founded in 1970 and remains active today, is also part of a larger organizational structure, but with a much more defined structure than Transition. The AG has elected leaders on its ExCom, formalized voting procedures, and committees such as a Conservation Committee suggested by the national Sierra Club. The Sierra Club is a mainstream environmental organization that attracts politically liberal participants. It engages in a range of tactics, including demonstrations and coalition work, but has certain regular activities such as tabling, petitioning, and electoral work on the campaigns of endorsed candidates.

Of the three groups that are not part of larger organizational structures, the Shadbush Environmental Justice Collective, which met from 2010 to 2014, was a radical environmental group with a collective structure. The group held frequent meetings and continually reflected on its strategic choices and their outcomes. Shadbush engaged in educational work to promote its radical perspective and efforts to help communities affected by fracking as well as demonstrations and other direct actions, often in coalition with other groups. Marcellus Protest, which met regularly from 2010 to 2013, was a diverse group of people with different backgrounds and ideologies with a shared opposition to fracking. The group structure included general assemblies, working groups, and a coordinating committee with fairly frequent turnover in membership. Its strategies and tactics included demonstrations, educational work, and efforts to ban fracking in Pittsburgh and other communities. Protect Our Parks, which met from 2013 to 2016, was a coalition campaign to stop the Allegheny County executive from selling the drilling rights for county parks. The group did not take a stand on fracking more generally but opposed fracking of the parks as a way of building a broad coalition of county residents. The coalition brought together activists from a variety of organizations, including the AG,

MP, and Shadbush, and it engaged in a range of activities, including testimony to the county council, lobbying, petitioning, and demonstrations.

CONCLUSION

The descriptions of the groups that I studied intensively reveal extensive activity by grassroots environmentalists in Pittsburgh as well as the considerable challenges that they faced. Most volunteer groups have trouble maintaining themselves over time, and this was the case with Transition, MP, Shadbush, and POP, which all ceased meeting by the end of 2016.[10] The Allegheny Group of the Sierra Club continues to meet and engage in its regular activities. This is not surprising insofar as the AG is the only group in my sample that is part of a federated organization with a clear structure and strategic agenda. Sierra Club staff members, employed by the national or state organizations, provide some assistance to local Sierra Club groups, though the AG does not have its own staff. Transition Pittsburgh was also part of a national and international organization, and the group got some advice from national organizers and a strategic model from the international organization, but there is much less connection among organizational levels in Transition than in the Sierra Club. All of the groups in my sample struggled to recruit members and to engage in ongoing actions and campaigns, but each faced different challenges. At the same time, the local movement accomplished a great deal during the period of my study in terms of mobilizing participants for collective action and raising public awareness about issues such as fracking and air and water pollution.

In the following chapters, I further examine movement mobilization by showing how these groups created organizational structures, devised strategies and tactics, and acted together. Not all of my cases are equally useful in examining these processes, and I do not attempt to use the cases in a completely parallel fashion. That is, I do not make systematic comparisons across cases, but employ the cases where they are useful for understanding movement dynamics. Transition Pittsburgh is different in many ways from the other more political groups in my sample and, as a result, is discussed less frequently, though the case remains useful in understanding some dynamics. My focus is on explaining the processes of movement mobilization and collective action, and I do not claim that groups in Pittsburgh are representative of groups elsewhere or even that the groups I have studied are representative of other

[10] MP continued to exist as a network after it stopped holding regular meetings when core members became involved in POP in 2013, and some MP/POP members got together for a few meetings in 2019. As described in Chapter 6, many individuals continued to be active after organizations stopped meeting.

groups in Pittsburgh. Rather, my goal is to use these cases to explain the dynamics of mobilization and problems that theorists know to be common to many grassroots movement organizations. My contribution is to provide a close-up view of the actual functioning of environmental groups based on years of participant observation in a variety of organizations.

4

Environmental Issues and Framing Work

Grassroots environmentalism depends on the passion and commitment of local activists, most of them volunteers, who donate their time, skills, and other resources to efforts to protect their neighborhoods and communities as well as public lands. It also requires some type of organization to bring community members together, as well as strategies and tactics, both to accomplish goals and to build the movement by engaging supporters in collective action. This chapter begins to explain the mobilization of grassroots environmentalism by showing how compelling environmental issues create concerns among local residents and lead many to become active. It also describes how movement actors work to frame issues in ways that capture the experiences and grievances of potential constituents. Chapters 5 and 6 then continue the story of how grassroots environmentalists mobilize by looking at how concerned individuals become movement participants and how movement structures and communities continue to organize them. To explain the process of mobilization, I draw extensively on my observations of the anti-fracking activities and campaigns that took place during the time of my study.

Many of the organizational and strategic challenges facing environmental activists are not unique to the environmental movement. Indeed, many activists with experience in other social movements bring their skills and multi-issue concerns to their environmental activism. There are, however, several ways in which environmental issues per se have affected grassroots organizing. One difficulty is that environmental problems such as invisible pollutants and global warming are sometimes seen as abstract and difficult for people to connect to their everyday lives. Particularly in the case of climate change, the "extended time horizon" associated with the issue makes it difficult to arouse concerns, even in the face of increasing extreme weather events (McAdam 2017:201–2). As Norgaard (2011:3) found, even people who were informed and concerned about climate change tended to focus on "more local, manageable topics" relevant to their everyday lives. Yet many environmental problems do directly threaten individuals and their families. The immediate impacts of environmental problems have motivated the participation of many people in affected communities, and the growing impacts of climate change can be expected to prompt even more involvement. Although many people in communities where environmental and health impacts have occurred continue to support fossil fuel industries, practices such as shale gas drilling have aroused a great deal of political opposition and community conflict (Dokshin 2016; Gullion 2015; Hochschild 2016; McAdam and Boudet 2012; Wilber 2012). In the Pittsburgh area, smoke and pollution from the burning of coal and from the steel industry have motivated environmental movements for decades (DeMarco 2017; Longhurst 2010; Tarr 2003).

Part of the work of attracting and maintaining participation in movements involves the channeling of emotions connected to threats and risks (Gould 2009; Summers-Effler 2010). Framing by environmentalists has informed people about the urgency of issues and aroused emotions. Documentaries, books, and other educational materials have motivated many supporters, and while such materials alone do not spark movements (Meyer and Rohlinger 2012), activists often use them to mobilize support. Visuals in documentaries such as *Gasland* (described below) have been extremely important in framing environmental issues as urgent. Moreover, participants in environmental organizations have engaged in their own eye-opening research on the issues, frequently using the internet. The experience of environmental problems, and much of the framing of issues, occurs at the community level; many activists either live in communities with environmental and health problems or come into contact with others who are directly impacted by environmental degradation.

MOBILIZING ISSUES

As noted in Chapter 2, threats often provoke collective action (Almeida 2019), but there is no guarantee that even cataclysmic changes such as global warming will spur movement participation. To understand why some issues mobilize people at particular times, we need to look at how people become engaged with issues as well as their meaning. The history of grassroots environmentalism in Pittsburgh and Pennsylvania shows that many individuals without prior movement experience become activists in the face of threats to their air, water, health, and way of life. Experienced environmentalists often remain involved in the movement for many years because they care deeply about pressing issues such as pollution and climate change. The following account shows how some key issues have mobilized both new and experienced activists in recent years.

Protecting Public Lands and Wildlife

Like the supporters of conservation organizations such as the Izaak Walton League and the Wilderness Society, many environmentalists who joined the Allegheny Group of the Sierra Club in the early 1970s enjoyed outdoor recreation such as hiking and camping. Some were attracted to the Club by its then extensive outings program and quickly became concerned about environmental degradation of public lands. For example, Peter Wray, a long-time member of the Allegheny Group, was typical of many early Sierra Club members. He recalled in an interview that his initial motivation for joining in 1974 was his desire to go on the outings with his sons, then aged 11 and 13, who enjoyed hiking:

> We then got involved in going on hikes and doing some backpacking and camping and pretty soon we realized the areas that we were hiking and camping in were often state forests, as well as the Allegheny National Forest. And appreciation of those areas led to a need for protecting those areas, because there were issues at that time, such as clear cutting . . . and fairly large construction [plans] for large lodges in some of the state parks and also in Allegheny National Forest. So that's how we got involved and then quickly things led from one to the other. (Interview, January 17, 2013)

It was not the case that Wray and others lacked interest in environmental issues before joining the Sierra Club. Like many others in Pittsburgh, Wray was concerned about air pollution, and he had previously joined the Group Against Smog and Pollution (GASP). Sierra Club membership led to his increased awareness of a wide variety of environmental issues. Wray started out as an outings leader in the Allegheny Group and became a member of the Executive Committee (ExCom) in 1975 before doing a couple of stints as the ExCom Chair. He recalled that the outings were the group's primary recruiting mechanism for many years and that "most of the people on the Executive Committee at that time, if not all of them, were outing leaders" (interview, January 17, 2013). Bruce Sundquist, one of the first members of the Allegheny Group (who died in 2016), organized members to map out trails in various state and national forests in Pennsylvania, publishing a series of trail guides that became an important source of income for the Allegheny Group as well as a valuable resource for the outings program (Wray 2016). Many people who experienced the outdoors and were attached to the wilderness areas of Pennsylvania became passionate about protecting them.

The outings program of the Allegheny Group eventually declined due in part to competition from other organizations, such as Venture Outdoors, a nonprofit organization that began leading similar outings in the 2000s, and Meetup, an online tool that began to be used to organize outings. With the demise of its outings program, the Allegheny Group lost an extremely effective way of recruiting activists. The group's Public Lands Action Team did take a few outings to the Allegheny National Forest to observe its natural springs and synchronous fireflies in the forest. The Pennsylvania Chapter of the Sierra Club also arranged outings in conjunction with its annual meetings, typically held in or near state forests or parks, and hired a staff member to support group outings programs. In 2018, the state chapter held training sessions for outings leaders, including one in Pittsburgh, and the AG recruited a young member to head a new outings committee in an effort to revive the program (fieldnotes, June 11, 2018). The route to membership through outings may, however, have inadvertently increased homogeneity in the Sierra Club. Reflecting on a finding by the National Park Service that 90 percent of visitors to national parks are white, the editors of the Pennsylvania Chapter newsletter commented: "Hiking and camping seem to be a white pastime. This may be one reason that the Sierra Club has had trouble diversifying its membership. Many people say they were introduced to the Club by hiking with one of its groups." The editors expressed hope that new Club programs aimed at reaching out to diverse constituents "will find new ways to help people of color fall in love with the many wild places in Pennsylvania" (Taylor and Coleman 2016:2).[1]

Concern about the management of public lands and the protection of wilderness areas motivated many Sierra Club activists. Samuel Hays, then an environmental historian at the University of Pittsburgh, was one of the leaders of a campaign to prevent a lodge from being built on public lands, as well as a drive in 1984 to include additional Allegheny National Forest lands in the National Wilderness Preservation System (Hays 2003; Wray 2010). Other early activities included efforts to oppose strip mining, which threatened water supplies, and to protect the Clarion River from pulp mill wastes (www.sierraclub.org). Nor did the Allegheny Group confine itself to state and local issues regarding public lands; as part of a national organization, the group became involved in numerous national campaigns over the years. For example, the AG was an active member of a coalition that lobbied President Carter to sign legislation protecting wilderness areas in Alaska in 1980 (Wray 2010). The AG's Endangered Species Action Team (ESAT) worked on a number of national issues such as lobbying for protection of the Arctic National Wildlife Refuge and petitioning for the placement of animals on the endangered species list due to threats from global warming. ESAT was concerned with the fate of popular endangered species such as polar bears and whales but also worked to protect lesser known species in Pennsylvania.

Activists worried about the protection of wildlife and public lands have long been concerned with the environmental impacts of coal mining in Pennsylvania. More recently, environmental groups in the state, including Sierra Club activists concerned with public lands and endangered species, have increasingly turned their attention to the threat of fracking as well. PennEnvironment

[1] Taylor (2016) details the history of racial and class exclusion from state and national parks and forests. While blacks and other people of color helped build and maintain park facilities, segregationists attempted to bar them from many opportunities for outdoor recreation. Nevertheless, Taylor shows that blacks, like whites, were drawn to opportunities to escape cities and visit wilderness areas. Historically, conservationists and preservationists either endorsed or ignored discriminatory policies, making it difficult for people of color and the working class to participate in environmental activities (Taylor 2016:382). Finney (2014:69) argues that the lack of visual images and textual representations of African Americans in the media, National Park Service brochures, and elsewhere made blacks invisible in the context of the "environment."

conducted extensive research on the issue and released a study entitled "Frack Attack" that detailed threats to parks, forests, and public lands in Pennsylvania (PennEnvironment 2014). The Pennsylvania Chapter of the Sierra Club worked in coalition with PennEnvironment, PennFuture, and other groups to safeguard public lands, such as Loyalsock State Forest, from fracking. More radical groups such as Marcellus Shale Earth First also engaged in "tree sits" and blockades to prevent pipeline construction and destruction of public lands and forests (stateimpact.npr, March 21, 2014; marcellusearthfirst.wordpress.com). Pittsburgh activists have supported various state campaigns in Pennsylvania, and the Protect Our Parks coalition came out of a larger concern with the infringement of fossil fuel industries on public lands.

Coal Mining

Numerous environmental groups have worked to mitigate the impacts of coal mining on the environment in Pittsburgh and Western Pennsylvania as well as in other regions of the country. The Izaak Walton League, which was founded in 1922, has been involved in efforts to protect Pennsylvania land and water from acidic drainage from coal mines (Casner 2003). In Pittsburgh, a smoke control movement addressed the problem of dense smoke from the burning of bituminous coal for domestic and commercial heating. While heavy smoke decreased with the adoption of gas heating and a municipal smoke control policy, air pollution in Pittsburgh continued to be a problem (Mershon and Tarr 2003). GASP participated in public hearings and other tactics in the struggle against air pollution in the city (Longhurst 2010). Environmental groups in Pennsylvania continue to monitor numerous problems associated with coal mining, including threats created by abandoned coal mines to streams and rivers in Pennsylvania, the impacts of toxic coal ash dumps on communities, respiratory problems from coal dust and other health problems for residents, and the need for remediation of public lands.

The Sierra Club has challenged destructive coal mining practices such as strip mining for decades. In recent years, through its Beyond Coal campaign, the Sierra Club has worked to retire coal plants and replace them with renewable energy, such as wind and solar, and to educate the public about the health and environmental impacts of coal mining and burning. The Sierra Club and the Center for Coalfield Justice (CCJ) have worked together closely, and members of the Allegheny Group living in rural areas outside Pittsburgh have worked with the CCJ and Beyond Coal staff on campaigns related to coal mining. Ken Yonek, a member of the AG who was also on the board of the CCJ, was prompted by a CCJ strategy session to think about how to increase contact with diverse communities in the Monongahela River Valley area. He contacted the local National Association for the Advancement of Colored People (NAACP) and was invited to a meeting and then asked to form an environmental justice committee for the branch, which he did, to work on the many issues facing low-income Mon Valley communities, including health and environmental problems associated with idled coal-burning power plants, coal waste disposal sites, abandoned mine drainage sites, and other conditions (Yonek 2015).

The state chapter of the Sierra Club and the CCJ teamed up to successfully challenge a permit issued by the state Department of Environmental Protection to allow more longwall coal mining near Ryerson Station State Park (Legere 2017a) and later worked together to appeal permit revisions granted by the DEP (Graham 2018). Despite their inability to stop longwall mining in the park, the CCJ and its allies have campaigned to force the industry to restore the park and to prevent further damages in addition to waging other campaigns to support communities affected by coal mining. In doing so, they have framed the issues in terms of community rights. As CCJ director Veronica Coptis noted in a profile in the *New Yorker*: "This is property owned by every resident in Pennsylvania. They don't get to keep plowing through our communities as if we didn't matter" (Griswold 2017:50).

Shale Gas Drilling

While environmental groups in Pennsylvania continue to deal with problems associated with coal mining, they have also become very involved in addressing the issue of shale gas drilling, which has motivated grassroots activism in shale gas regions across the country as well as in the state. Many residents of affected areas joined local movements or took action against the industry as "accidental activists" (Wilber 2012) or "reluctant activists" (Gullion 2015) when they experienced disruption of their homes, damage to their land, contamination of drinking water, and severe health impacts associated with shale gas drilling (Griswold 2018). Others, including seasoned environmentalists, became alerted to the dangers of fracking and organized to stop its spread and to support those in affected communities. Shale gas drilling began in the Barnett Shale of northern Texas in the early 2000s, with much of the drilling taking place in urban locations in the Dallas–Fort Worth region (Gullion 2015:39). Residents, most of whom had never before participated in collective action, were completely unprepared for the invasion of their middle-class neighborhoods but felt compelled to become involved to protect their families and communities (Gullion 2015). Shale gas drilling operations expanded to numerous other states, including Colorado, Arkansas, and Louisiana, and opponents mobilized around concerns such as water contamination (Ladd 2018).

In the Marcellus Shale region, which includes large areas of Pennsylvania and New York as well as parts of Ohio and West Virginia, numerous "landmen" began to arrive around 2006 to convince residents in New York and Pennsylvania to lease their land for gas drilling in the Marcellus Shale. By 2008, New York had declared a moratorium on fracking, while the "gas rush" in rural northeastern Pennsylvania was in full swing (Wilber 2012). Dimock,

Pennsylvania became a hot spot in the controversy over fracking as the previously quiet area became an industrial site, and concerns about truck accidents, disposal of waste water, gas leaks, methane migration, and well contamination accumulated. By 2009, extensive well water problems were apparent in Dimock (Wilber 2012:92), and problems in other shale drilling communities were later made public (Nobel 2017; Urbina 2011a).

Awareness of the dangers of fracking increased as experts began to debate the merits of unconventional gas drilling and as citizens affected by fracking began to speak out about its impacts. Among academic experts, a professor of Geosciences at Pennsylvania State University, Terry Engelder, strongly promoted shale gas drilling; in 2008, he calculated that the Marcellus Shale contained enough natural gas to meet US energy needs for decades, and he made himself available as a source for reporters covering the issue (Wilber 2012:2–3). On the other side of the debate, one of the most influential experts was Anthony Ingraffea, an engineering professor at Cornell University who began to publicly challenge industry claims in 2009 after witnessing inaccurate public relations campaigns (Wilber 2012:150). *Gasland*, the 2010 documentary by Josh Fox, was extraordinarily important in raising consciousness about the dangers of fracking and influencing public discourse around the practice (Vasi et al. 2015). In this documentary, filmmaker Josh Fox narrates the story of how his family was offered $100,000 to lease their land in the Delaware River Basin area of Pennsylvania for shale gas drilling, prompting Fox to investigate the experiences of people living in communities across the country affected by fracking. Fox tells the stories of many families who experienced health problems they attributed to fracking, and the film contains dramatic images such as tap water that could be set on fire due to natural gas contamination. Individuals living in affected communities also shaped the debate by speaking out about the dangers of fracking – testifying about air pollution, water contamination, unbearable noise, loss of property values, the deaths of their animals, and human health problems such as rashes, stomach aches, nausea, asthma, difficulty breathing, severe nosebleeds, and tumors that they attributed to fracking (see *Shalefield Stories* 2012, 2015).

In Pittsburgh, concerns about fracking began to emerge in 2009 and 2010 as awareness of the drilling activity in Pennsylvania and nationwide was spreading. Many people who had never previously been involved in local environmental struggles became involved in the movement against fracking in Southwestern Pennsylvania and Pittsburgh as well as in other regions and states (Gullion 2015; Wilber 2012). Threats from fracking outraged a number of residents after they were approached by landmen to sell the drilling rights to their property. To become involved in the opposition to fracking, they needed pathways to the movement, and the internet was an important source of information. Many people found their way into the movement by connecting with others concerned about the issue, including some experienced activists. The Pittsburgh anti-fracking movement gained strength from both newcomers to environmentalism and long-time activists, who helped to organize events and create an infrastructure for the movement.

Many early participants in Marcellus Protest came from areas threatened by fracking. In Butler County, about an hour north of Pittsburgh, landmen were visiting residents and encouraging them to sign leases. Some began signing, but others started doing their own research on the internet. Diane Sipe, a resident of Butler County who worked in Pittsburgh at the time, went to the internet after she and her husband were visited by a landman and wondered whether or not they should sign a lease. She recalled their search:

> The first thing you go to, at least back then, the first thing when you would Google "Marcellus Shale," you got the Penn State [Terry Engelder study]. So you read that, you would think [it's a good thing]. This is what happened – that's why, people who signed back at that time, I don't really judge them harshly. They didn't know, and if they had done that, and they got to that, I mean, that would be perfectly reasonable if you read all that stuff and said, oh, okay, this is okay. You know, because at that time, why would you not trust the Penn State study? But it just so happened we kept going and we came across Josh Fox's website. The movie [*Gasland*] hadn't been released yet, but he had a website, and we started looking ... oh, my gosh, look at some of this stuff. I think there was the water flaming on there. Anyhow, what we concluded is that the gas has been there a really long time and it's not going anywhere and there's no reason to rush into this. And if it turns out to be a good thing, you know, we can do it, and hey, maybe even the price is better. So anyhow, we didn't sign the lease. (Interview, May 9, 2012)

In the city of Pittsburgh, home owners were also being approached by landmen. Mark Schneider was a resident of Lincoln Place and a Pittsburgh fire fighter who had previously served in the army for ten years. An outspoken person who was not afraid of confrontation, Mark said of his opponents: "They had no idea they were stepping into a bee's nest with me. They had no idea" (interview, May 15, 2012). He became incensed about fracking and, along with his wife Elizabeth and some of their neighbors, spearheaded the early opposition and formed the Lincoln Place Action Group. Mark recalled his reaction to an offer received by email:

> I said, if it sounds too good to be true, it is. So I went in the other room in here [at his home] and I Googled it. And Googled it. And Googled it. And Googled it. It was like Alice in Wonderland. The more I learned, the deeper the rabbit hole got. The deeper it got, the darker it got. And the more I would learn – and I learned a lot – the more I learned, the more I started going, wait a minute, not in my country, not in my city, not in my neighborhood. (Interview, May 15, 2012)

He was further outraged when he attended a one-sided community meeting organized by the company that was signing leases. However, Mark and Elizabeth met a few other residents opposed to the drilling at the event, and they also met an experienced environmentalist who told Elizabeth about Stephanie Hallowich, a woman who was speaking out about the health and environmental impacts of fracking on her family and farm near Hickory, Pennsylvania. At the same time, through his internet searches, Mark learned about Ron Gulla, a farmer in the same area, who had leased his land in 2002, but later spoke out about the damage done to his land and animals by fracking. After listening to Gulla's YouTube videos, Mark contacted him by email before speaking to him by phone. Mark and Elizabeth and three of their neighbors later drove to Hickory to see for themselves the impact of fracking. They met with both Ron Gulla and Stephanie Hallowich and toured their farms. Elizabeth recalled that this experience was critical to their understanding of the impacts of fracking:

> I felt very strong about it after seeing it firsthand at Stephanie's and knowing what happened to them and what's happening to all the people and the people in Dimock and the things that are occurring in other parts of the country. I'm just like, how could it be that all these things have been happening for ten years, at least ten years, and nobody's looking into this? Nobody's making the connection. And to see – not that we are the only thing that started it here, because I know that we're not. I know if we didn't, someone else would have. And the connection now that they're finally making with health effects and all these other, you know, groups that have started ... (Interview, May 15, 2012)

Appalled by what they saw and heard, the Lincoln Place residents were determined not to allow fracking in their neighborhood or city. With guidance from the experienced environmentalist they met, their group passed out fliers and organized a large town meeting at Mifflin School in Lincoln Place with a panel consisting of Bob Donnan, Ron Gulla, Stephanie Hallowich,[2] and others, including a scientist from the University of Pittsburgh and a staff member from Clean Water Action. They invited their city councilor, Doug Shields, who was just learning about the issue and who would later sponsor Pittsburgh's ban on fracking. Mark Schneider and others were interviewed on a local radio show, where they discussed various issues related to fracking, including water use and disposal and the contamination of drinking water. Many people called in with questions from different areas in the region, and after this publicity, people began to contact the Lincoln Place Action Group about the issue. Fracking was becoming a new topic of public discourse, and more people were becoming concerned about it.

[2] Donnan has spoken publicly about his experiences with fracking and posts pictures and commentary on his blog (see www.donnan.com/blog). Ron Gulla has also spoken publicly and recorded multiple videos and commentaries about his experiences with fracking (www.marcellus-shale.us/Ron-Gulla) Stephanie Hallowich spoke publicly and was interviewed on videos for a variety of environmental groups until she and her family finally reached a settlement with the gas company Range Resources, which included a controversial gag order (see https://en.wikipedia.org/wiki/Stephanie_Hallowich). The court records were later ordered unsealed after a legal challenge by the *Pittsburgh Post-Gazette* and Hallowich was permitted to testify under subpoena (Hopey and Templeton 2019).

As residents of areas targeted by the industry investigated the issue and learned of the harm being done to people in areas of the state where fracking was already occurring, they became increasingly alarmed. Ken and Loretta Weir, who became outspoken opponents of fracking, were among the residents from Lincoln Place who attended the meeting at Mifflin School after getting a flyer in their door. They were outraged by the stories of the people already affected by fracking and the lack of protection for citizens and the environment. Loretta recalled:

They had quite a crowd there that night. So we went to that meeting and I was outraged. Honestly, they had me in at the Halliburton loophole. I was like, there's something very wrong here. Then I hear about the Halliburton loophole, the exemptions from major provisions of the Clean Water Act, Clean Air Act, you know, the superfund law, and it just didn't take much more than that. I mean, I was really upset about it ... This whole issue for some reason, there's just a time in your life where something just really draws you in. And when I looked at that issue and heard those people that were affected, and we really started to look at this and started to look things up, you know, on the computer. It was unbelievable. (Interview, April 19, 2012)

Many Pittsburgh activists developed close relationships with people harmed by fracking in areas across Pennsylvania and in some cases around the country. While movements do not always organize in places where energy projects create harms (McAdam and Boudet 2012), here we see citizens actively generating concerns about the effects of fracking. The Lincoln Place Action Group made a video about fracking featuring some of the people who were harmed. Members of the Shadbush collective were also concerned from the start to tell the stories of people in affected communities, and they interviewed some residents of Southwestern Pennsylvania for a video project aired in episodes on Rust Belt Radio, the local Indymedia station. A number of activists from Marcellus Protest and other groups worked together in a group called Friends of the Harmed to compile the testimonies of people affected by fracking in two volumes of *Shalefield Stories* (2012, 2015). These publications are glossy and professional looking, having been written, edited, and designed by skilled volunteers and printed with donations from Steel Valley Printers, Mountain Watershed Association, and other individuals and organizations. They are also powerful in conveying the experiences of ordinary people, in Pennsylvania and in other parts of the country, with fracking.

Many of the testimonials talk about living in the countryside of Pennsylvania or other states, enjoying the peace and quiet and the clean air and water. They describe how this all changed as the industry moved in; in the words of a Fayette County, Pennsylvania resident: "The truck traffic is constant. Land damages were of course immediate. Trucks, noise, dust, and nomadic workers followed. Then came the pollution of our air and water, then deforestation and the destruction of fruit trees ..." (*Shalefield Stories* 2012:15). Beyond the noise and destruction of the environment, many people describe the serious health problems that affected their families and killed their animals. They also told of the contamination of their water and the lack of response by both the industry and the DEP, which some residents dubbed "Don't Expect Protection." A resident of the Woodlands area of Butler County describes how her water turned black and her hair began to fall out, along with muscle pain and other health problems. After the gas company had her water tested and found unacceptable levels of arsenic and other contaminants, she and other neighbors were provided with water buffalos until the DEP determined that the water problems were not due to drilling (*Shalefield Stories* 2015:20–2). In another testimonial, a truck driver from Cambria County, Pennsylvania, told of becoming disabled after hauling and emptying wastewater from fracking without adequate protection (*Shalefield Stories* 2015:15). These and many other stories of life in the shalefields were extremely compelling, motivating many activists who wanted to assist people harmed by fracking.

Climate Change and Community Sustainability

The climate crisis is now the overriding environmental issue, and despite difficulties in creating a grassroots climate movement (McAdam 2017), the issue arouses a great deal of passion among committed environmentalists. In the Allegheny Group of the Sierra Club, a Global Warming Action Team (GWAT) was organized in 2009, and its members worked on issues such as fracking and opposition to the Keystone XL Pipeline, which were seen as major contributors to global warming. Allegheny Group members organized buses and traveled to Washington, DC in February 2013 for a demonstration, organized by the Sierra Club, 350.org, and others, focused on climate change and stopping the Keystone XL Pipeline. Along with members of Marcellus Protest, Shadbush, and others in the Pittsburgh environmental movement community, they also helped organize buses and traveled to New York City to participate in the People's Climate March in September 2014. In April 2017, Sierra Club members joined other Pittsburgh environmentalists in traveling to Washington, DC for the second People's Climate March, which marked 100 days of resistance to the Trump administration. In Pittsburgh, hundreds of environmentalists, including a contingent from the Sierra Club, participated in the local March for Science, held on Earth Day, April 22, 2017, and the People's Climate March on April 29, 2017.

These demonstrations were extremely inspiring to grassroots environmentalists, and the experience helped create solidarity and ongoing collective action (cf. Meyer and Kimeldorf 2015). The long bus rides to Washington, DC and New York provided an opportunity for some participants to speak together about the issues. One of the results of the 2013 and 2014 events was that a group of local activists began to have conversations about how to address climate change more directly, eventually forming a steering committee that started a Pittsburgh chapter of 350.org. The issue was clearly an extremely pressing one for many activists concerned about issues of social justice and the impacts of global warming on future generations. At the launch of Pittsburgh 350.org, for example, long-time Sierra Club activist Peter Wray spoke emotionally about how, at age 81, he didn't expect to see the worst consequences of climate change but that he feared his grandchildren will inherit a very different world than the one we live in now, and that climate activism is for all of our children and grandchildren (fieldnotes, April 1, 2015). Patricia DeMarco, another Sierra Club member, has also spoken and written passionately about "the ethical challenges of climate change" and its impacts on our children and grandchildren, animals, and people in less-developed countries who are likely to experience unfairly more of the harms of climate change even though they did less to cause it than those in developed countries (DeMarco 2017).

For many grassroots environmentalists, the issue of climate change points to the need for major changes in the culture and economy. As a collective with a radical, multi-issue approach, Shadbush emphasized economic and environmental justice; while focusing on the issue of fracking because of its importance to Pennsylvania, the collective saw the issue as part of "a broader movement for climate justice" (Shalefield Justice Action Camp program, November 2012). In addressing the issue of global warming, radical environmentalists argue for a "just transition" away from fossil fuels that will not put the burden on disadvantaged people and communities of color. In the fall of 2011, after the Occupy movement began, Marcellus Protest and Shadbush labeled gas industry operatives part of "the 1%" and called for "economic and environmental justice," "community self-determination," and a "just and sustainable future" as we transition away from fossil fuels and "develop energy resources and infrastructure that are renewable, safe, and locally self-sufficient" (flyer for "The People's Reception," November 2011).

Other local groups also called for environmental and climate justice. The Environmental Justice Committee, which was created in September 2012 as part of the Thomas Merton Center (TMC), a Pittsburgh peace and justice center, considered the dominant economic model unsustainable and called for environmental justice and community control (Environmental Justice Committee flyer, 2012). Transition Pittsburgh was also concerned with community-level solutions to global warming and other environmental problems. Transition's approach was very attractive insofar as the group framed these problems as opportunities for local empowerment; despite the need for communities to "power down," they could build resilience and a better way of life without excessive consumerism and energy use. Based on the international Transition model, local activists were excited by the idea that they might tackle climate change and decrease reliance on fossil fuels while building community through projects such as permaculture, community gardens, and skill sharing.

In October 2017, when Al Gore's Climate Reality Project came to Pittsburgh to conduct a training of "climate reality leaders" to equip local activists to educate their communities about climate change, more than 1,300 people from around the world attended (Hopey 2017a). The workshop was positively reviewed by the many Pittsburgh activists in attendance, who have since continued to work on educating people about climate change and advancing solutions. A participant wrote about her experience in the Pittsburgh 350.org e-newsletter:

> I had the opportunity to meet many other Pittsburghers who are doing the work, and made plans to partner with some to do Acts of Leadership together. AND I shared information about 350 Pittsburgh and hope that means we will get some new members. And ... the training was wonderful. I couldn't imagine sitting for three days with only brief breaks without going stir-crazy but I did, and never felt bored or distracted. The training inspired me to go deeper and further in the fight against what is our pollution-based economy, powered by fossil fuels. (Issue #39 of 350 Pittsburgh's biweekly digest, October 26, 2017)

Thus, climate change continues to motivate participation in the local environmental movement as activists connect it to other problems with pollution caused by the extraction and burning of fossil fuels.

FRAMING AND NAMING

Collective action framing connects the concerns of individuals to those of organizations and the larger movement and helps draw people into movements (Benford and Snow 2000). As described earlier, many people became opposed to fracking after learning about people harmed by the industry, and they sought to tell their stories. Activists are generally aware of the importance of framing issues for media and public consumption and as a way of attracting members, but the process of framing is often difficult. Even the naming of groups and events, which is one type framing work, can be demanding. Naming requires participants to think about the impact of the name on potential members, as well as other targeted groups, and to come to some consensus about what the group stands for. Often there are practicalities involved in framing and naming, such as what fits on a bumper sticker or yard sign and the difficulties of "rebranding" after a group has an established identity. Even when a group agrees on messaging, it is impossible for a movement to control the ways in which individuals discuss issues; democratic movements cannot maintain such control over participants even if they want to do so.

Early framing of the fracking issue helped the movement reach a variety of constituents. In addition to concerns about human health and environmental damage, fracking was framed in terms of community rights. As the movement ramped up, many activists cited an article of the Pennsylvania Constitution, known as the Environmental Rights Amendment, that was added in 1971:

> The people have a right to clean air, pure water, and to the preservation of the natural, scenic, historic and esthetic values of the environment. Pennsylvania's public natural resources are the common property of all the people, including generations yet to come.

As trustee of these resources, the Commonwealth shall conserve and maintain them for the benefit of all the people. (Article I, Section 27, PA Constitution)

This article, a legacy of the resurgence of environmentalism in the 1960s, was very inspiring to activists, and it was frequently quoted on websites and blogs, in movement literature, and on t-shirts and signs. Moreover, the frame was linked to an important tactic advanced by an existing organization. The Community Environmental Legal Defense Fund (CELDF), based in Mercersburg, Pennsylvania, used a rights frame in drafting Pittsburgh's ban on fracking as a community rights ordinance. CELDF stresses community rights and the rights of nature in its work to help communities assert their rights to local self-government and to fight against corporations in banning practices such as fracking, factory farming, and water privatization (http://celdf.org). This approach was important in launching the anti-fracking movement as CELDF worked with Pittsburgh city counselor Doug Shields to write the Pittsburgh ordinance, gave workshops to activists, and helped other municipalities write community rights ordinances.

The community rights frame helped unify Marcellus Protest, which began as a diverse group of people without a real consensus on goals. This lack of consensus was evident at an early meeting of the first Outreach committee, formed before the November 3, 2010 demonstration, where we discussed the need for a statement of our position to put into a letter soliciting endorsements. Several people noted that there are a range of positions, including a complete ban on drilling, a moratorium, regulation or oversight, or taxing. A member of the Lincoln Place Action Group said she wasn't even sure where she stood – originally, she was in favor of regulation, but she was now leaning more toward a total ban. We agreed that Outreach should be an umbrella group that supported various forms of opposition to shale gas drilling. We did not want to create divisions among supporters, though the group did want to come up with a statement about what we stood for (fieldnotes, September 10, 2010). At a meeting of the larger group, Marcellus Protest tackled the problem of inclusiveness in its effort to come up with a mission statement. Because the discussion took place before the November 3 march, some participants argued for inclusive language to attract people to the protest. One said, "We need every damn body on November 3." Another argued, "We need an alliance of groups with different levels of understanding. It's okay to be a little watered down when you're a coalition." After much discussion, with some initially favoring a stronger statement that the mission was to "stop drilling," the group settled on a less direct statement that read: "Marcellus Shale Protest is an alliance of Western PA groups and individuals building a broad movement to stop the destruction of our environment and communities caused by Marcellus Shale drilling" (fieldnotes, September 24, 2010).

There was no discussion of the group's name at the early general assemblies, and "Marcellus Shale Protest" was apparently adopted by the ad hoc group that called the public meetings to organize the November 3 protest. Prior to the first of these meetings on September 3, 2010, David Meieran set up the website using the marcellusprotest.org URL, and in doing so, he cleverly mimicked elements of the website of the industry group, called the Marcellus Shale Coalition, including its logo. Marcellus Shale Protest was a counter to the industry name that was later informally shortened to Marcellus Protest. As the group moved from organizing a protest to creating a movement organization, the original name continued to be used, but participants began to talk about changing it. One reason for the dissatisfaction was that the name did not seem like that of an organization and did not reflect the group's broader focus beyond organizing protests. Another problem was raised by the Outreach committee; particularly for those working in conservative rural communities, the word "protest" was not helpful. Outreach members worried about using literature that referred to protest and directing people to the Marcellus Protest website when they were just learning about the fracking issue and might be turned off by what they perceived as a radical group.

These concerns resulted in extensive discussion between Outreach and David Meieran, the webmaster, as well as with other members of Marcellus Protest. One possible solution discussed at length was a "dual-domain" approach to the website in which a different URL could be used on Outreach literature to direct users to the same website without duplicating work by creating a whole new website. Although this solution was acceptable to Outreach, it was never implemented, perhaps because the group got bogged down in a broader discussion of whether the name Marcellus Protest should be changed. In addition to the concerns raised by Outreach about reaching rural and conservative residents, group members were worried about how the name might affect fundraising efforts such as applications to foundations. David designed a survey to ask if members thought the name should be changed, suggesting some possibilities as well as allowing respondents to propose names and asking them about their involvement in the group. After several weeks, he reported on March 1, 2011 that 80 percent of fifty-nine respondents favored a name change. In addition to the survey, there was much online discussion of a possible name change and of the desirability of using "protest" in the name.

Even though a large majority of MP members favored a name change, it turned out to be very difficult to accomplish, with a lot of different concerns expressed. One was that "rebranding" would be difficult, given the group's considerable following and the popularity of the website in Pennsylvania and beyond. Members also debated whether or not the word "protest" really turned people off. Some were concerned that alternative names, such as Marcellus Community Alliance or Marcellus Action Alliance, which were two options on the survey, could easily be confused with the names of pro-industry groups. One person bitterly suggested "Marcellus Education Outlet" for those who wanted to "distance yourselves from being 'against' something" while warning that this and other names suggested might "attract a lot of pro-drilling folks." Others tried to sidestep this criticism by coming up with suggestions that would make the group's stance clear. For example, one contributor proposed following the lead of MP's fiscal sponsor at the time, the Center for Coalfield Justice, by adopting a name such as "Communities for Shalefield Justice." Another problem, however, was that it was impossible to make a decision, particularly through email debate. This led one person on the email list to suggest a face-to-face discussion about the direction of the group because he thought disagreement over the name reflected significant political and strategic differences within MP. He added, "I can't say I have ever seen a broad strategic discussion go well over e-mail."

Although Marcellus Protest did have subsequent discussions about strategy and the direction of the group (see Chapters 7 and 8), the decision about a name change was dropped. There are clearly practical problems with a name change after a movement organization has been active for a time, though of course some groups do successfully change their names. It may be more difficult to change the name – and direction – of a relatively diverse group in which there is a lack of consensus. Moreover, as discussed in Chapter 2, the structures of groups create different advantages and disadvantages; in this case, the decentralized structure of MP allowed subgroups to make their own choices. During the discussion of names, Loretta Weir suggested "Communities United for Rights and Environment" and later used that name for a group she formed to create media ads opposed to fracking. Members of Outreach from Butler County created their own group with its own website, called Marcellus Outreach Butler (MOB), while maintaining ties to MP. Over time, MOB became less cautious in its positions as it gained some strength, though it was still strategic in its approach within a conservative community. Diane Sipe noted:

> I remember us debating whether to have a ban or a moratorium and we took the more conservative moratorium. We've changed our mission statement to we want a ban, you know, and we're less reluctant to state our position. And we've had a protest ourselves, you know. I mean, I don't think I'd still want it in our name . . . I wouldn't want protest part of our name. (Interview, May 9, 2012)

Thus, groups do make changes in their frames and positions, but they are influenced by their political environment as well as by internal group dynamics.

Framing decisions are also influenced by the initiatives of individuals and by practicalities, making it difficult to control framing. For example, when the MP Coordinating Committee talked about wanting bumper stickers, one person said that Dana Dolney had already created "family friendly" bumper stickers with the message "I [heart] Pennsylvania . . . don't frack it up! marcellusprotest.org" and a graphic of a sun over mountains. The intention of having something "family friendly" had never been discussed by the group, though some disputed the characterization in any case and at least one person worried that the use of "frack" could offend some people. A member of Outreach was working on another design with the message "fracking is toxic" and a poison symbol on it after the idea was approved by the Outreach committee, but it was never completed. It was easier to use the design that was already done, which most people seemed to like well enough.

When discussions about messaging occurred, participants were influenced by media coverage, allies, and opponents, as well as practical considerations. In the case of yard signs, the message "Fracking poisons air and water. Fight back now. Marcellusprotest.org" was approved after discussion by the Coordinating Committee. In considering the message, we decided that it should include both air and water, noting that a recent *New York Times* story discussed the release of gas into the air as well as wastewater problems associated with fracking (Urbina 2011b). Mel commented approvingly that this would make it relevant to the local GASP group and also that the industry hates the word "fracking," so we should definitely use it (fieldnotes, April 12, 2011). John found a union printer for the signs, while Dana got a contact of hers to create the design. The yard signs paid for themselves as people purchased them, and when more were needed, it was easy to call up the supplier and order them. It would not have been easy to change the message or design; after one member later criticized the signs as "too busy looking," there was no enthusiasm for creating new signs when there were so many other things that needed doing. Both the bumper stickers and yard signs took individual initiative, and their creators moved on to other tasks.

Marcellus Protest also undertook more extended efforts at effective framing with its strategic messaging committee. Aaron Booz, who was particularly interested in framing the issue to combat messages from the gas industry, initiated and led this effort. After he called an initial open meeting, a small group of interested and like-minded people began meeting regularly with the goal of coming up with effective ways of talking about fracking. The committee proceeded by reading materials about messaging and the objectives to consider in devising frames. We talked about many issues, such as how to construct frames that appeal to different audiences, how to counter industry frames, how to address issues exploited by opponents such as the need for jobs, and how to present our values and concerns about fracking. We spent a great deal of time considering the appeal and effectiveness of various frames for different audiences and ultimately produced a "toolkit" that was shared on the website to guide activists in framing the issue in an honest and convincing manner. The toolkit promoted frames based on positive values, including preservation of the environment for future generations, community self-determination, health and safety, environmental justice, and sustainability. It also identified threats to those values, such as greed and consumerism, the failure of government to represent citizens rather than corporations, and industrial destruction of the environment. Besides the half dozen people who were regulars on the committee, however, it was never clear how many people took strategic messaging seriously, and it was impossible to control framing by individuals and groups in the movement, effective or not.

In my interviews, some people said they did look at the toolkit and thought it was useful; others said they hadn't thought much about messaging. When asked about strategic messaging, Mel said: "I haven't thought a whole lot about that. It was still just packaging the issue of our future and our children's future and the safety of our air and water and land. That still has to be stressed over and over and over again" (interview, June 19, 2012). Wanda said she had used the strategic messaging toolkit and thought it was very helpful (interview, July 27, 2012). John said of the effort at strategic messaging: "I think my mind set is probably a little bit more fatalistic, you know, that there is no magic formula that's going to bring people around ... there is no magic phrasing that people are going to say, oh, now I get it, if they don't want to get it ... I don't mean that it's not good ... You can always do it [messaging] better" (interview, May 2, 2012). Claudia said she read the toolkit carefully and did incorporate some ideas from it into the public speaking she was doing for Outreach. She added, however, that it is so hard to get other people to volunteer to speak that she couldn't imagine adding in expectations about framing. Regarding speaking engagements, she said, "I just wish we had someone who would do it at all" (interview, May 2, 2012).

While MP members expressed mixed views on the importance of strategic messaging, Shadbush organizers agreed on the importance of framing and naming. Framing or messaging was often mentioned at meetings, and the importance of "telling our story" was discussed at retreats. After discussion at the first retreat in 2012, Shadbush held an internal workshop on messaging at which Patrick and Ben led us in discussing the "Battle of the Story" (www.storybasedstrategy.org/battle-of-the-story) framework and applying it to Shadbush's work (fieldnotes, June 22, 2012). Everyone in attendance seemed pleased with the exercise, but it was never explicitly employed in subsequent work as far as I know. At the strategic planning workshop in March 2013, open to outsiders, another Shadbush member gave a presentation on the importance of telling our story and conveying our values and evoking emotions through narrative. Again, it was not clear how this information was used in the subsequent campaign planning, but inclusion of this discussion demonstrated Shadbush's commitment to strategic framing, and some individuals may have used the information in their work.

This commitment was evident from the start in Shadbush's choice of name. The group was very deliberate about naming themselves and thought about a name that would reflect what they were trying to accomplish. Jessica, who contributed the Shadbush name, recalled:

> We had a number of different names, and I came up with that one because it kind of exemplified a lot of connections that we wanted to make with our group. Like the shadbush is a native tree; it grows wild in the forest and it was used by Native Americans for food. But it also is planted a lot in the city and grows in the city and people can, you know, find it and eat it here and use it as a landscaping tree. It's appreciated. I'm not sure how many people actually know what it is, but it's in our environment. And we wanted

that connection between the city and rural areas to be something that our group helped create and the very name of the bush itself – it's called the shadbush because it blooms as the shad run. The shad is an anadromous fish ... it swims up the river to spawn and then goes back out. And so like for someone to have named that, they would have known that and the shad would have been important to them ... We're important to them, right? And so that kind of connection, like the material dependence and the awareness of the elements of our natural world was, you know, something that we wanted to foster as well. (Interview, May 4, 2015)

The group did consider naming themselves after the hellbender, a type of salamander, or the black locust, a tree, but decided against these choices. Patrick liked these ideas, but bowed to the group consensus, and good naturedly joked about this on a few occasions. When the group assessed the action with the giant pig, Jean Marie argued that it was good for media coverage because it was "not scary or angry," and another member said the focus on supporting local agriculture was helpful as "good positive framing." Casey added that their message was both "militant" and "strong," and Patrick said he thought "we hit a sweet spot with the tone" of the action. Patrick also mentioned the Shadbush name as a positive element, saying it would not have been as good if it had been the "black locus collective" or the "hellbender collective." He explained with humor to those not involved in Shabush at the start: "We had a protracted debate about the name of the collective. I lost!" (fieldnotes, February 10, 2013). When I asked him about the name in an interview, Patrick explained:

We didn't have a name for a long time and ... I think the idea was that we would have some sort of plant or animal that was kind of indigenous to our bioregion ... so we kicked around a lot of different ideas ... and people thought that the Black Locust Collective and the Hellbender Collective – both of which would have been my preferred options – sounded too, you know, confrontational, so we ended up going with the Shadbush Collective. (Interview, April 12, 2015)

Jessica similarly recalled:

We talked about [the hellbender] at the time ... I think it seemed hardcore, it seemed militant, you know, and we wanted to be a group that did direct action and stuff, but we debated it and ... I think [another group member] looked at a hellbender and she said, "Oh, that's scary." And who knows what one is anyway? And it sounds pretty metal – and black locust did as well. And from the very beginning we didn't want to be like some radical punk collective off in their own, you know, personal and aesthetic world. We wanted to be a little bit more relatable. So people were like, yeah, okay, we'll go with Shadbush even though they kind of wanted the other thing! (Interview, May 4, 2015)

The Shadbush name provided an identity of sorts, both inside and outside the collective. In addition to frequent references to one another as "berries," Shadbush members sometimes talked, perhaps not completely in jest, about getting some type of shadbush tattoos. At one meeting, Ben brought a pitcher of smoothies made from shadbush berries, which he had picked from bushes in the area. When I commented that I didn't know you could eat the berries, Jessica said, "Oh yeah, that's the point, it's a useful berry" (fieldnotes, June 23, 2013). Beyond the group, Shadbush was becoming known as a result of successful activities such as the action camp and the pig action. The collective was being asked to lend its name in support of many activities, particularly among radical environmental groups in the region. As Patrick commented when the group was evaluating the success of the action camp:

> It's a double-edged sword. We got street cred and there are going to be more requests to sign on and support things. It will matter more if we are promoting an activity; there will be expectations. This is the first big thing we've done. We need to think about stewarding the good reputation we've built and our credibility. And we have to assess our capacity and say no sometimes. (Fieldnotes, October 20, 2012)

As thoughtful as Shadbush was about naming and framing, the group could not anticipate all of the issues that might arise. Ben told me in an interview that he initially liked the name, but changed his view over time:

> I think we all kind of connected with the metaphor of shadbush as both a native plant and as a plant that is cultivated and grown in the city and that sort of metaphor of trying to work across the divide between urban and rural ... I came to dislike the name because ... it was kind of like mysterious who we were, and the name didn't say very much about who we were and what we did. And I think that was a barrier to reaching out to people, to bringing new people in ... we weren't really reaching very far beyond our own kind of subculture. (Interview, April 29, 2015)

Even though Shadbush had built up "street cred" and had a good reputation among radical environmentalists in the region, when the group disbanded and a new group formed, participants chose to rename themselves Three Rivers Rising Tide (TRRT). While some Shadbush members were sorry that the group name would be changed, others felt it was necessary for the group to have a new identity in order to expand. Patrick argued that, while Shadbush had a regional reputation, the new group gained "a name that has national credibility" through its connection with Rising Tide North America (interview, April 12, 2015).[3]

Naming and framing have important consequences for attracting members and public support. In the case of Protect Our Parks, the campaign planning group that created the coalition started out by framing their concerns in terms of the impact of fracking on public lands. When the Allegheny County executive threatened the county parks with his plan to sell the rights to drill under Deer Lakes Park, the group decided to respond with a campaign around the parks. In advance of its second meeting, one member suggested that the group call itself the County Parks Defense Campaign. However, Aaron argued that

[3] As it turned out, TRRT dissolved after a few years; while the new identity and name change expanded the group's constituency for a time, it was not enough to create a stable organization.

the phrase "our parks" should be in the name because he thought it was important for conveying the sentiment that the parks belonged to us. After briefly debating the merits of "Defend Our Parks" versus "Protect Our Parks," we decided on the latter, which had a nice acronym and alliteration. The frame that the county parks belong to the citizens of Allegheny County was indeed a powerful one, and many outraged people outside the core POP group volunteered to collect signatures on petitions and to testify at county council meetings against the drilling plans. Opponents of drilling received a great deal of publicity in the local media, increasing the visibility of the movement despite the inability to stop fracking at Deer Lakes Park.

CONCLUSION

The history of the Pittsburgh environmental movement shows that area residents have cared deeply about numerous "burning issues" over the years. Compelling environmental problems are of great concern to many individuals, some with personal grievances, and they motivate the participation of many people in grassroots organizations and communities. However, there is also active work involved in connecting people concerned about issues to movement organizations and activities, and activists need to frame their concerns convincingly to build a movement. Movement organizers have made use of educational materials and frames that attract a variety of potential constituents, and individuals have also sought out information. Once individuals become involved in environmental activities, they often expand their awareness of issues and motivations for participation (Snow et al. 1980). In the case of fracking, the threat brought new constituents into a local movement that had long battled the pollution caused by coal and other industry in Pennsylvania. Internet communications allowed people who became alarmed about fracking to find information and to locate others concerned about the issue. Framing of the issue by pre-existing organizations such as CELDF and new organizations such as Marcellus Protest and Shadbush was critical to the launching of the local movement.

This chapter showed how the local movement engaged in framing on the issue of fracking and some of the challenges involved in this work. One challenge is the need to bridge differences among grassroots environmentalists with different backgrounds and ideologies. Participants in emergent organizations such as MP had to carefully hammer out positions, as they did in the case of the original mission statement, to satisfy constituents with different goals and backgrounds. Another challenge is to develop frames that address the lack of racial diversity in much of the environmental movement (Finney 2014). Recognizing this problem, members of organizations such as the Sierra Club have at times reached out to other organizations such as the NAACP and consequently adopted new frames such as environmental justice. Coalition work is an important source of frame bridging and diffusion. When groups and individuals interact with different types of constituents, they may adopt new ways of thinking about issues. However, new framing challenges, which groups are not always equipped to address, emerge over time. Difficulties involving the term "protest" in Marcellus Protest came about not primarily because of differences among activists, but because residents of rural and urban communities were working in different political environments. Offshoots such as MOB found their own solutions to framing problems such as creating their own websites.

Framing work was affected by group structures as well as by the backgrounds and ideologies of participants. In an established organization such as the Sierra Club, much of the framing work is done by professional leaders at the national level. However, local groups may also participate in coalitions that expand their framing. For example, the Allegheny Group of the Sierra Club is a member of Pittsburgh United, a progressive alliance of community, labor, faith, and environmental organizations with professional staff, which coordinates a variety of campaigns, many focusing on the rights of workers. The AG has been active in Pittsburgh United's Our Water Campaign, which employs an environmental justice frame and works to attract diverse leaders and constituents (http://pittsburghunited.org/ourwater/). In a decentralized group like MP, framing is more difficult to accomplish and control. Marcellus Protest relied heavily on internet communications with its sprawling and amorphous membership to supplement face-to-face meetings. Consequently, MP was unable to come to consensus on a name change that might have helped the group present itself to the public. Shadbush was a much more ideologically cohesive group, with only minor framing disputes, and the entire collective met frequently to hash out issues. Even in this group, however, there may have been unforeseen consequences of framing choices, such as difficulty in attracting more members.

Framing requires coordination and initiative to produce effective messaging. Some cultural frames, such as the community and environmental rights frame, predate movement campaigns, but they nevertheless require individual or organizational entrepreneurs to introduce them to new activists and emerging organizations. In the case of Marcellus Protest, quite a few activists were aware of the Environmental Rights Amendment in the Pennsylvania constitution, and they used it repeatedly in organizing work. CELDF and other organizers also promoted a community rights frame in drafting the Pittsburgh ordinance and in workshops to spread this approach to other areas. Other frames are deliberate products of activist work. Although MP was a diverse group, one member initiated a strategic messaging committee that helped create a framing strategy. Nevertheless, that committee had no control over how others in the diverse group might use its work and frame the issue.

In short, environmental problems confronting people in their communities and the motivating frames devised by activists helped mobilize the grassroots environmental movement in Pittsburgh, though not without difficulties. As we have seen, the challenges of framing work are related to the diversity and structures of groups, as well as practicalities and unforeseen circumstances. The work of framing is ongoing and requires continuous evaluation. Beyond issues and frames, Chapters 5 and 6 show that ongoing movement mobilization requires the engagement of individuals and strategies and tactics that encourage ongoing involvement, effective organizational structures, and the creation of relationships and networks in the movement community.

5

Mobilizing a Grassroots Environmental Movement

Nationwide, a broad range of environmental issues such as air and water pollution, threats to public lands, endangered species, environmental justice, longwall coal mining, mountaintop removal, community sustainability, public transportation, fracking, and climate change have motivated grassroots involvement. The movement has mobilized a variety of constituents, including retirees, students and other young people, religious groups, residents of affected areas, labor unionists, and long-time progressive activists. In Pittsburgh, the movement includes established organizations such as the Group against Smog and Pollution (GASP), Clean Water Action, PennEnvironment, and the Sierra Club, as well as newer environmental justice, climate action, and anti-fracking groups. Grassroots activists have employed a variety of tactics, from conventional lobbying and electoral politics to demonstrations and direct actions. They have organized numerous campaigns and participated in many alliances. The movement has frequently succeeded in engaging a large number of people, though many potential supporters have not become involved, and it is difficult to maintain organizations and collective action for an extended period of time. The movement has won some important battles, but continually struggles to make real progress in protecting the environment.

While compelling environmental issues alarm potential movement constituents, ongoing mobilization requires strategies and tactics to build the movement and accomplish goals. Because movement momentum can easily decline, movements need to keep recruiting new participants and devising new actions and campaigns in order to grow and succeed. This requires the development of solidarity and collective identity among participants and the creation of networks and organizational structures that involve activists. Supporters need to learn tools of involvement in collective action and public life and to find the types of organizations and activities in which they feel comfortable participating. Relationships among activists in movement organizations and communities are critical to the continuing involvement of individuals. Movement organizations need to facilitate the design and implementation of new strategies and tactics, and alliances and cooperative relations among movement organizations are required for movements to carry out large actions and ongoing campaigns.

This chapter focuses on how individuals became participants in existing and emergent movement organizations and on the role of tactics in mobilizing them. Chapter 6 continues to explain the process of mobilization by looking at the movement infrastructure that was created, including networks in the movement community as well as movement organizations.

PARTICIPATING IN THE MOVEMENT

As discussed in Chapter 4, grassroots activists are motivated by their concern about a wide range of environmental problems. The movement has often used persuasive frames in appealing to people with strong feelings about the issues, which are likely to accompany personal experiences and visible problems. But issues, frames, and grievances are not enough to mobilize a movement, and even critical events such as environmental disasters do not automatically create protest (Staggenborg 1993). Grassroots environmentalism expands when local activists find effective ways to organize and devise strategies and tactics that bring in new participants and build the movement. While grievances and ideas motivate some people to seek out environmental activism, many are drawn into activism by movement organizations and their campaigns and tactics. Once involved, they may develop ties to other participants and learn more about the issues, thereby increasing their motivations and concerns.

The structures and cultures of movement organizations and communities are critical to activist engagement. For individuals to become participants, they need to come into contact with the movement, either through pre-existing organizations such as the Sierra Club or through an emerging movement such as the anti-fracking movement. Once they make contact, they need to figure out how to become involved. When motivated individuals feel included and useful to movement activities and form relationships with others within organizations and the movement community, they are likely to become strongly committed. Some movement organizations are better than others at encouraging participation, and the available forums for participation vary across organizations and over time. As noted in earlier chapters, diversity is a challenge for many groups, with regard to both attracting and managing members of organizations and forming alliances. Bridge leaders, and practices that unite diverse constituents, are often needed to provide ways for participants with different backgrounds and ideologies to become involved in a movement. Strategies and tactics are critical in this regard; groups need both mobilizing strategies to recruit newcomers and tactics to engage participants in ongoing movement campaigns.

In the following sections, I describe the experiences of individuals who became involved in different types of grassroots environmental group. I begin with a look at how some individuals who were concerned about environmental problems found their way into an established organization, the Allegheny Group of the Sierra Club. I then describe how some people became involved in the emerging movement against fracking. As we will see, the differing structures and cultures of the organizations affected their ability to attract participants, but they experienced some similar problems in that campaigns and tactics were needed to attract more participants, while more participants were needed to mount campaigns and devise tactics.

Joining the Sierra Club

The Allegheny Group of the Sierra Club is important as a stable organizational structure, where individuals can find environmental activism in Pittsburgh. The AG regularly participates in national and state Sierra Club campaigns and also regularly works in coalition with state and local organizations such as the Center for Coalfield Justice, Clean Water Action, PennEnvironment, and GASP. As a member of Pittsburgh United, the AG participates in campaigns such as the Clean Rivers Campaign and the Our Water Campaign (later combined as Our Water, Our Rivers). The AG has a core of active members who engage in ongoing activities such as political campaign work and phone banking to mobilize activists for various events and demonstrations in addition to contributing to coalition work. National Sierra Club initiatives, and staff located in Pittsburgh, have played an important role in recruiting some of the group's core activists. Whereas many early activists were recruited through the outings program, a number of newer members became active during the 2004 election campaign. The national organization increased its efforts to organize grassroots supporters with its Building Environmental Communities campaign, and Pittsburgh was one of the cities where the Club opened a field office in 2004 and placed staff, who helped organize local members. Many local members of the Sierra Club who were not yet active were contacted by email or phone. As Barbara Grover recalled:

> This email comes out, the timing is perfect for me because I'm retiring, okay, "we're looking for people, we're going to have a general meeting, we're going to become politically active in preventing the re-election of Mr. George W. Bush." Well I'm right there, that man, he was driving me nuts ... I credit George W. Bush [with recruitment]. We all showed up at that meeting and they started, you know, it was "this is what we're going to do. We're going to start to do phone banking, we're going to do activities, we're going to have a Tuesday night activist night. If you can, come down here." (Interview, December 13, 2012)

A group of people, who remained active for years to come, joined in the effort to prevent George W. Bush's reelection. Although that endeavor was unsuccessful, it brought new activists into the AG, and staff members helped to involve them in statewide campaigns. As Barb explained, with the guidance of staff and experienced volunteers, she became more and more involved in the group:

> I was just a volunteer. Then, oh at some point, you know, there'd be a rally and they would say, "Barb, would you speak at this rally?" You know because I used to tell them this is the best thing that happened to me since sliced bread because I was looking for something that in my retirement I would be contributing in some way to society. I can't sit at home and just do nothing ... I was so enthusiastic, this is the best thing, I was so thrilled because we were actually accomplishing something, we had a mercury campaign, I couldn't believe it. Anyway, so then they said oh would you speak at this place, so then would you speak at this and why don't you take on this role, I mean it was ... and then Earth Day would come along and we would canvas. I was learning all this stuff, I had no clue as to how to canvas or phone bank I mean, you know, you just do all that stuff and they [said], you know, well why don't you come and do this. Earth Day stuff whatever, okay, we'll canvas, so what am I supposed to do? So I learned how to do that and there were people who did it with me who had been doing this for 30 years and then it's, well, why don't you run for the ExCom ... You know, and why don't you – we're going to have this group of people, you know how to canvas so why don't you be a leader and explain it to these people. It's step by step [to becoming more involved]. (Interview, December 13, 2012)

Thus, campaigns such as the Sierra Club's effort to get the state of Pennsylvania to impose stricter rules on mercury emissions helped integrate new volunteers into the organization by involving them in activities such as gathering signatures in support of the legislation (Lesle 2006). Importantly, members formed relationships with one another and a commitment to the Sierra Club. When asked how he became committed to the Sierra Club, Mike, who joined at the same time that Barb did, said, "I would just say that it was basically the relationships that were built. I mean, if the first group that had given me a call was Greenpeace and I made friends that way, then I'd be volunteering for Greenpeace ... You know, it's not particularly the organization per se" (interview, May 2, 2013). Naomi, who also became involved around the same time, said it was a combination of getting to know other activists and becoming increasingly interested in the issues, particularly climate change. As she explained, "I just kept going just because I met people through it and I wanted to save more and more of the planet and I really like got into the issues" (interview, May 5, 2013). Chris, who joined a few years later in 2008, recalled that he was looking for an environmental group to volunteer with because he was concerned about climate change. He happened to encounter the Sierra Club at a table in Schenley Plaza staffed by volunteers, who invited him to join the group. He said, "It was a nice group of people," and they drew him in with an event that was part of the Cool Cities Campaign, a bus tour of the city focused on solar panels and green technologies, followed by a press conference. Asked what kept him involved, he replied: "the relationships with other people, I think. I enjoy coming on Tuesday nights and talking to my friends" (interview, May 2, 2013). In addition to activist nights on Tuesdays, the AG held an annual holiday party, which was an opportunity to socialize and invite others in the community to mingle with Club members. For several years, the AG also organized a summer picnic, but discontinued this when attendance dropped off, and it was decided that it was too much work to organize both a picnic and holiday party.

Within the AG, some participants not only formed satisfying relationships but also learned strategies and tactics that seemed worthwhile. After creating the Endangered Species Action Team (ESAT) in 2009, members who were passionate about endangered species engaged in both lobbying and educational tactics, which increased their commitment. They won an early victory by helping to win either "endangered" or "threatened" status for four species of Allegheny River mussels with their testimony before the Pennsylvania Fish and Boat Commission (Allegheny Group of the Sierra Club 2010:3). This victory was important in energizing the team and was mentioned on numerous occasions at meetings and strategy sessions as an accomplishment; ESAT members had testified at a legislative meeting, and one team member reported, "They told us to come back more often" (fieldnotes, January 12, 2014). The team continued to engage in lobbying and educational work on behalf of all species but, in particular, "troubled native Pennsylvania species such as bats, mussels, raptors, and snakes" (Chute 2012:6).

ESAT also developed partnerships with the Pittsburgh Zoo and the city's Environmental Charter School and ran an active educational program aimed at youth. These activities helped ESAT to come up with effective messages; as one member said at a strategy session, "we should not forget the power of children." Children at the Environmental Charter School made posters and put their faces on mussels, she said, and "the legislature remembers things like that" (fieldnotes, January 12, 2012). ESAT members often described their activities, such as tabling at festivals, as fun ones in which they interacted with children and adults, educating them about issues and collecting names on petitions and postcards (Chute 2012:7).

Members developed a collective identity; they seemed proud of their affiliation with the Sierra Club, and there was often a sense of being part of a larger organizational network of like-minded people. Nevertheless, the AG struggled to attract more active members. Despite participating in some active coalitions, and achieving some legislative victories along with its allies, much of the AG's recent work, while worthwhile, did not bring in a lot of new supporters. For example, members of the Global Warming Action Team regularly tabled at events such as neighborhood festivals, where they distributed literature and asked people to sign petitions such as one in opposition to the Keystone XL Pipeline. After the team was no longer active as such, its members continued to participate in weekly action nights at which they called members and asked them to write or call legislators, help table at events, attend public meetings, or work on political campaigns. While these activities could be enjoyable, they also involved hard work that often had no visible payoffs. Existing members were friendly people who liked one another and felt committed to the group, and this commitment along with regular meetings kept them coming back. Although they were welcoming to new people, the activities they engaged in were not always exciting (e.g., making phone calls), and persons who were not already friends with group members did not always return after a single time. The group seemed to have a hard time breaking out of its standard repertoire of actions. Outsiders or newcomers are often a source of fresh ideas (Polletta 1999, 2002), and the difficulty in expanding tactics could have been due in part to the lack of new participants who might urge regular AG members to do so.

After the election of Donald Trump, however, this began changing as the AG participated in resistance events such as the climate march and attracted some new members to the group. Sierra Club staff also reported receiving numerous expressions of interest from people who wanted to become active in the movement (fieldnotes, January 9, 2017). In April 2017, the state Sierra Club chapter hired a new staff member, Eva Resnick-Day, who began actively organizing Pittsburgh residents, including many young people, as part of the Sierra Club's "Ready for 100" campaign to get cities to commit to transitioning to 100 percent renewable energy by 2035. As she reported, "There was an outpouring of brand new activists looking to get involved after the People's Climate March ..." (Resnick-Day 2017:6). Randy Francisco, a national staff member, also began organizing volunteers to work on state and federal lobbying activities in 2017. While members of the AG were not all on these staff-led teams, the staff members regularly met with the ExCom chair to share information. Sierra Club staff play an important role in creating ties among different groups and activists in the movement community.

In addition to connecting with other activists, volunteers need to find meaningful ways to participate in groups in order to become committed to them. Many volunteer opportunities are of the "plug-in" type, which involve limited time and commitment (Lichterman 2005:58). Such opportunities for involvement may be offered by "low-engagement" groups that focus on turning numbers of people out for events or getting them to sign petitions, for example, rather than developing volunteer skill and commitment (Han 2014). However, they may also be seen by "high-engagement" groups as a way to gradually increase the involvement of newcomers. In the case of the AG, the group did want more committed activists, and some volunteers and staff argued that it is best to draw people in with simple requests at first and then build relationships. This worked in the case of activists such as Barb and Mike, who became more and more involved over time. However, the committed core of AG activists remained small during the period of my observations, and this may have been due in part to a belief among some members that most people do not want extensive involvement. For example, ESAT was a tight group of less than ten active members with an online membership list of about 100 people. While the online members were clearly interested in supporting the team in some way,

ESAT did not put its energies into developing their commitments. One active member explained at a strategic planning meeting that ESAT always assumes that people have no time and money, so "we give actions to them, like we ask them to sign petitions" (fieldnotes, December 13, 2014).

The AG did succeed in attracting some new members, both to work on action nights and to serve on the ExCom. Some of these newcomers were recruited by other volunteers and some by staff members paid by the national or state Sierra Club. Randy Francisco often supervised interns who he then encouraged to join the AG. One former intern, a college student, became an enthusiastic member who was elected to the ExCom (but later had to resign when she moved out of town). Randy also trained some group members in one-on-one recruitment techniques, which involve talking to prospective members to see what they are interested in, finding ways for them to get involved, and building relationships. However, the group had trouble finding enough people to do the one-on-one contacts and also faced the problem of not having enough different things for prospective members to do. Allegheny Group members recognized that more varied activities are needed to recruit new volunteers, but they suffered from the ongoing problem of needing more activists before they could undertake new activities and mount exciting campaigns that would attract more people. In other words, campaigns are needed to attract volunteers, but volunteers are needed to run campaigns.

Another problem in the Sierra Club was the lack of a seamless structure of teamwork between staff and volunteers. Staff and volunteers did help one another out, and they had good relationships with one another, but their work was often separate or parallel rather than tightly coordinated. Randy was basically helping the AG with recruitment in a volunteer capacity, as this was not part of his regular job as a national staff member, and he had limited time to contribute. Two staff members employed by the Pennsylvania chapter of the Sierra Club, Tom Hoffman and Eva Resnick-Day, worked to draw volunteers into their campaigns. Staff organizing helped the AG by providing new ways for members to become involved and also by encouraging new volunteers to join the group. However, the AG itself had no staff and no systematic ways of recruiting and integrating volunteers, such as a recruitment committee.

Becoming Active against Fracking

The shock associated with fracking, and the early excitement and passion generated by the movement, helped the newly emerging anti-fracking movement attract many participants. The early organizing against fracking often felt spontaneous to participants; as Mark Schneider commented, "This was like a vortex ... it was happening to us. It wasn't like we were activating; we were being activated" (interview, May 15, 2012). Not only was the issue a powerful one that could be further researched on the internet but individuals who were becoming outraged about fracking were also coming into contact with one another. Ken and Loretta Weir did not know anyone in the movement previously, but they started to meet people at the early meetings. Soon afterward, they attended another meeting organized by concerned residents in New Kensington, where they met more of the people who became part of the emerging movement against fracking. Diane Sipe and her husband also attended that meeting. Mel Packer, a long-time Pittsburgh activist who was collecting signatures at the time to get on the ballot as a Green Party candidate for the Senate, attended the early meetings in Lincoln Place and New Kensington. He recalled being "very impressed with the organizing and impressed with the people that showed up at these things. You know, just average working-class folks who decided they needed to take things into their own hands and weren't being fooled by politicians ... They already knew that they weren't going to get much help from those politicians" (interview, June 19, 2012).

At the same time that neighborhood groups like the Lincoln Place Action Group were organizing, individuals were learning about the issue through media publicity and the film *Gasland*. Many early activists were recruited after viewing the documentary, including some who already had environmental sympathies, but who were just becoming aware of the fracking issue. Some learned about the issue and later experienced threats in their neighborhoods, which motivated them to become heavily involved in the movement. One person who became active in Marcellus Protest (MP) and helped found the neighborhood group South Hills Area Against Dangerous Drilling (SHAADD) saw *Gasland* when it aired on HBO in June 2010. Recalling her reaction to the documentary, she said: "I panicked. I literally have been panicked ever since then ... I was really upset" (interview, June 26, 2012). Another SHAADD founder, Aaron Booz, read about the issue in the newspaper and saw *Gasland* at the Frick Park showing in August 2010. He later heard about a Marcellus Protest general assembly after listening to Rustbelt Radio, the local Indymedia station. He attended that meeting, the large one following the November 3 demonstration, and decided then that "I have to be involved in this. These people are passionate and they're doing stuff" (interview, June 26, 2012). After learning that four sites in the South Hills area where they lived were being zoned for fracking and drilling, the two residents met one another at a local council meeting and later organized community members through SHAADD.

While outrage about fracking and the excitement of a new movement motivated much of the early participation, it was later difficult to maintain that momentum. This was due in part to the decline in the threat felt by neighborhood groups, particularly after the Pittsburgh ban on fracking was won. Nevertheless, many activists continued their engagement in the movement through MP, Shadbush, and other organizations that worked to find ways to involve participants. Importantly, extensive networks and relationships were created, helping sustain the movement even when some individuals did not continue to attend meetings regularly. As new local groups such as Marcellus Outreach Butler (MOB) and SHAADD were organized, their leaders continued to participate in the movement and maintained close relations with the MP network. Ongoing actions and campaigns were critical to maintaining and building the movement by providing ways for individuals to become involved.

Marcellus Protest did, for a time, have ways to integrate new volunteers, but the opportunities for involvement were somewhat hit-or-miss, and newcomers typically had to show initiative to become involved in meaningful ways. In MP's first year, the Outreach committee provided some ways for enterprising individuals to participate. For example, Claudia Detwiler quickly found a niche for herself after coming to a meeting and signing on to help create a speaker's bureau. Although she found that it was hard to recruit others to speak, she and her husband John and a few other persons ended up filling a lot of requests for speakers. Claudia commented, "I was really glad to have come to the Outreach meetings because that's how I wound up with the speaker's bureau" (interview, May 2, 2012). Other people who went to Outreach meetings did not find roles easily, and it was common for newcomers to attend only one meeting. Many people who tried to get involved by contacting the group via a volunteer form on the website were unable to do so, even though the website managers passed on volunteer information to working groups. The member who received the website forms filled out by persons interested in Outreach said she didn't know what to tell them other than to come to the meetings. This problem was discussed at Outreach meetings but never resolved despite various ideas and plans. Dana suggested making lists of things we need, saying, "We're losing members because people can't get involved" (fieldnotes, February 16, 2011). Aaron noticed that half a dozen people who signed the form on the website volunteered to do media work. He suggested sending email suggestions about media stories they might comment on, which he did a few times (fieldnotes, April 28, 2011). The group also came up with a list of other things that new volunteers might do in preparation for a Marcellus Protest "meet and greet" event, such as letter writing and work on a media committee, but the event was not advertised well and attracted only a few potential recruits (fieldnotes, May 12, 2011). The Coordinating Committee intended to hold more such events but ended up lacking the capacity to do so. An ongoing problem for integrating newcomers was that existing volunteers were stretched too thin to do the work needed to recruit more people.

Even highly motivated persons wanting to become involved in MP did not always have an easy time finding activities they enjoyed and that used their skills. For example, Terri found that volunteers were needed for many tasks that required computer skills, which she lacked, making it difficult for her to find a niche in the organization. She said, "It's been hard ... I'm frustrated now and I'm feeling like I'm not doing anything" (interview, April 23, 2012). Unlike many people who failed to find roles easily, Terri persisted and later became heavily involved by joining the Coordinating Committee. As in the case of Outreach, committees often lacked the capacity to incorporate newcomers, and a member of the Coordinating Committee lamented in discussing the problem, "It is like we're attracting people to a spider web where they get stuck" (fieldnotes, September 1, 2011).

Shadbush, for its part, could be hard to join because its founders envisioned the group as a small collective that would screen potential members. Ben explained the original reasoning behind the process:

> Everyone who helped start Shadbush was from the start interested in working in this kind of semi-closed collective.... A big part of it I think was the desire to be working with people who you knew well and trusted and the idea that that would make us more effective. And also I think we had an idea in the beginning that we might be doing a lot of direct action and that it was important to have a small group in which we all knew each other very well. But in reality we didn't end up doing very much direct action and ... I think there was also just kind of a cliquey part of it that maybe we didn't examine very much or was inevitable ... You know, I think we knew each other well and liked each other and wanted to work together and we weren't thinking big about bringing people in. I think we had this idea that, or I had an idea that, we'll be a collective and we'll kind of work autonomously and find ways that we can partner with people on certain projects but not necessarily become a bigger group and with the thinking that this is a model and other people can start other collectives that can be working autonomously but together in certain ways. (Interview, April 29, 2015)

Consequently, Shadbush developed a process whereby someone interested in joining typically had an initial one-on-one meeting with a Shadbush member, who explained the group's procedures and perspective. Potential recruits who agreed with Shadbush's principles of unity were invited to attend a meeting before being accepted by the group as a member. In practice, this process was sometimes truncated, particularly when the prospect was known by multiple members of the group. A few persons who wanted to join were rejected by the group before attending a meeting (usually because they were thought to be ideologically unsuited or difficult to work with based on past experiences of members with the individual), and more prospective members lost interest when told of the process involved in joining.

Shadbush members ended up split on the usefulness of the group's model and process. Some of those who went through the process and joined Shadbush found it fairly easy in practice and felt that it served the purpose of ensuring that the group remained tightly knit and trusting of one another. Casey appreciated the process, saying, "In my experience, I think for small groups, it's really important that you be on the same page" (interview, April 3, 2015). Robin initially had a difficult time connecting with the group, even though she knew Jean Marie, a Shadbush member who was enthusiastic about recruiting her. When Shadbush invited outsiders to help edit the second edition of the *Shalefield Grassroots Reporter*, Robin volunteered but was never contacted to help, simply because the group lacked sufficient organization to incorporate everyone who volunteered. However, she ended up becoming a core member of Shadbush and felt that the collective model and the membership process were worthwhile. She explained: "I feel like we did have this kind of core that we felt very invested and took real responsibility, more than I've seen in some other groups, where it's sort of like I'll do it if I can or if I want to. And I think having that sort of process of like reaching out to people who we knew and trusted sort of helped that happen" (interview, April 9, 2015). Patrick, one of Shadbush's original members, noted both the benefit and the cost of the process: "When people would get in, we're able to hit the ground running, there's a level of familiarity, and I think that kind of having ideological organizational alignment was good, but then the flip side of that is that we also end up demographically homogenous" (interview, April 12, 2015). While the process was devised originally because the group expected to be doing a lot of direct action, the result, as Jessica commented, was that "we ended up not having enough members" (interview, May 4, 2015). Ben eventually concluded that the model of a "semiclosed collective" was ineffective and that it made recruitment very difficult:

> I think that [the model] was just unrealistic and I think ultimately it made us less effective, because I think we just became pretty insular. And as far as the actual membership process, what I remember about it is that we would meet with people or, you know, we'd find someone who would express interest and then we would say, okay, well we have to have a one-on-one meeting with one of us from Shadbush about going to a meeting with them to kind of feel them out, go back to the group, talk about it, make a decision, go back to them ... Invite them or not invite them, and that ... I mean I think that whole thing, there weren't that many people who actually got it enough to understand why we might be going through that. And I think that just created a barrier. I mean, I think there were plenty of people who expressed interest who we either never followed up with or we'd kind of explain the process and they're like, okay, well thanks anyway. (Interview, April 29, 2015)

While the decision to remain a "semiclosed collective" kept Shadbush small and limited its capacity, the group was quite active for a time and managed to pull off many activities by drawing on a much larger network of friends and members of the radical community in Pittsburgh. As Polletta (2002) finds, a group culture that includes an ethos of participatory democracy and strong ties among participants helps collectives to function despite risks of exclusivity. Shadbush had a strong culture of active participation, and I was continually amazed at how much work collective members routinely volunteered to take on. As described in Chapter 3, the collective staged a number of impressive actions with support from friends and allies. In discussing the structure of the collective, Robin commented on the importance of this support network and suggested that more structure may have been needed:

> We wouldn't have been able to do the things we did if we didn't have like a wider network. I think it may have been a good thing to have a structure more of like a coordinating committee and a way for that wider network to feel more connected and like they also were Shadbush. Because it sort of petered out, I feel like, because there was a small group that was so involved and then as people got more demanding jobs or got into other projects it just sort of petered out. (Interview, April 9, 2015)

As this comment suggests, it is difficult for a collective to continually replace key members when they leave as well as to integrate persons outside of the core. There were discussions within Shadbush about the possibility of formalizing its network and creating an outer tier of members who were not interested in coming to frequent meetings but wanted to be part of the group. Whereas the core Shadbush members often referred to themselves as "berries" (for the fruit that grows on the shadbush), they jokingly referred to the imagined outer tier as the "outer berries" or sometimes the "farm team." The group talked about having monthly potluck dinners, at which volunteer opportunities would be explained, creating an email list of supporters, and developing lists of people willing to do certain things (fieldnotes, July 8, 2012).

They did create an email list for sending announcements and soliciting help, and one potluck was organized for persons in the extended network as a first step toward creating a more formalized relationship. While the group intended to schedule regular potlucks, they recognized that this would be a fair amount of work and did not end up carrying out the plan. Moreover, people who came to the first potluck were somewhat confused as to what Shadbush wanted them to do and how they could get involved in the group. Collective members realized that they needed ongoing activities or campaigns to keep a network of supporters connected to the group, and they discussed various activities that "outer berries" might help with, such as organizing tours of areas where fracking was occurring, but they were not prepared to provide ongoing tasks for volunteers. Thus, they faced a similar dilemma to that of the Sierra Club insofar as it takes a fair amount of effort to bring in new people, even in a relatively minor capacity.

For those who joined the Shadbush collective, there were many ways to participate, and the group was highly dedicated to developing the skills and commitment of members. In its first year (2010–11), Shadbush members were heavily involved in researching the issue of fracking and putting together educational materials such as a "Fracking 101" presentation to explain the issue, an independent media project in collaboration with the Pittsburgh Independent Media Center featuring interviews with persons in communities affected by fracking, and the first issue of the *Shalefields Grassroots Reporter*, published in Summer 2011. When the collective decided to work on a second volume of the *Grassroots Reporter*, published in Spring 2013, the group formed an editorial committee and also solicited articles and editing assistance from others in the movement community. As Robin's experience showed, not all volunteers were contacted, but one member of MP who became involved in working on the paper later became an active member of Shadbush. Others who volunteered developed closer ties to Shadbush without joining the collective. There were also multiple opportunities to help organize activities such as the action camp, held in November 2012, and the various demonstrations and direct actions described in Chapter 3.

Shadbush was concerned that all members be able to develop skills such as facilitating meetings, dealing with the mass media, and organizing actions. The desire to build member skills and commitment was explicitly discussed at meetings. For example, the following discussion occurred at a Shadbush meeting:

> Robin said that she sometimes hesitates to volunteer for things because she thinks we want the "best job done" and she did not "want to do it and let people down." Jessica said when someone wants to volunteer but doesn't have experience we should have a conversation about what is necessary for the job and who can help; this is important for "capacity building." Casey said when people accept responsibility for something, they should voice their reasons, which might be wanting to learn a skill. Robin agreed it would be helpful to think you can volunteer for something when you don't yet have all the skills needed. (Fieldnotes, March 9, 2013)

One member of the collective, Patrick, was more experienced in direct action and radical politics than most other members. He played an important role in the group and had a big influence on it, but he also worked hard to teach skills to other members and encouraged others to take the lead when possible. Other members who had particular skills also made efforts to teach them to others in the group. As a result, participants in Shadbush reported that they learned a great deal from being members of the collective; Casey's comment that "I learned a lot about how to work in a group" (interview, April 3, 2015) was a common sentiment. Robin elaborated on how the "pig action" (described in Chapter 3) affected the group:

> I loved the pig action, just something about it. And I think that a lot of the positives that came from the pig action weren't like – it's not like we had a demand that got met. You know, we closed down the well path for the day, we got media attention and had messaging. But we had the goal to just sort of increase everyone's capacity and I think that happened in a really big way, and a lot of people were able to plug in and do roles they hadn't done before and participate in an action that they hadn't done before and that might intimidate them. And I think it was just a really awesome learning experience. I think we spent two months organizing that . . . Like most actions you don't spend that long organizing them but I think we spent that long because we were really structuring it as this sort of learning experience. (Interview, April 9, 2015)

Shadbush was clearly a "high engagement" group (Han 2014) that provided many opportunities for intensive involvement and skill development. Moreover, the group also involved members of the larger movement community in many actions, and those people also expanded their skills and commitments to the movement. As described in Chapter 3, members of the Butler County group MOB collaborated with Shadbush on a number of occasions and strong relationships developed among members of the two groups. At the debriefing for the pig action, Michael, one of the MOB members who had participated, commented about MOB's enthusiasm, saying that Shadbush "built capacity in the region with the action." He said that the action at Ellis's office with

Shadbush got MOB people excited, followed by the action camp, and then the pig action. Now, Michael said, they were talking about "what are we going to do next in Butler?" As the debriefing session was ending, he added: "Thank you for including me. This has been an awesome experience, and I felt close to everyone" (fieldnotes, February 10, 2013).

MOBILIZING TACTICS

As the above account of how people get involved in grassroots environmentalism demonstrates, collective action is essential to movement building. Awareness among members of the public about the impacts of an issue like fracking does not automatically produce a movement. As outlined in Chapter 2, both organization and strategy are critical to mobilization. Organizers need to find ways to activate people, including those who are already concerned as well as those who are not yet aware of a problem. Movement organizers employ mobilizing tactics such as program meetings as means of recruiting people and maintaining organizations (Reger and Staggenborg 2006). In this section, I focus on the experience of the Pittsburgh anti-fracking movement to show how mobilizing tactics were central to the growth of the movement.

The documentary film *Gasland* served as the basis for an important mobilizing tactic not only because its content was powerful but also because activists used it as an effective organizing tool. On June 5, 2010, many early movement participants joined hundreds of others at the Byham Theater in Pittsburgh for a free preview screening of *Gasland* with Josh Fox before the documentary aired on HBO later in the month. In August 2010, about 500 people attended another special screening of Gasland in Frick Park, with Fox again in attendance. Later, organizers of the emerging movement showed the film over and over in small venues and discussed it with their audiences. *Gasland* made many people who lived outside of affected or targeted areas aware of the issue for the first time, and quite a few of them were motivated to join the anti-fracking movement as a result. For some, the experience was profound; for example, a Sierra Club member commented in introducing herself at a public meeting that, after seeing *Gasland*, "my life was forever changed" (fieldnotes, December 4, 2010). In addition, Josh Fox himself had a big impact on the Pittsburgh movement as he visited the city multiple times and developed close relationships with some of the local activists. His message was alarming, but also hopeful. One activist commented: "Every time Josh Fox is giving something, I'll always listen to his talk. Because, for some reason, he's really like uplifting, and you're like, okay Josh, we can do this ..." (interview, January 18, 2011).

Many activists became involved in lobbying for the Pittsburgh ban, which was passed by the Pittsburgh City Council on November 16, 2010 (Smydo 2010). Residents of areas of the city where leases were being signed and neighborhood groups were being formed, including Lawrenceville and Lincoln Place, were quickly drawn into the citywide movement to lobby for the ban on fracking. Lincoln Place residents contacted their city councilor, Doug Shields, with their concerns about shale gas drilling and consequently he began to investigate the issue (interview, April 3, 2012). Lawrenceville residents also contacted their city councilor and became involved along with the Lincoln Place residents in the citywide effort. Gloria Forouzan recalled:

> I got involved with the issue in June 2010 when I first learned from my neighbors that drilling leases were being signed in my urban neighborhood of Lawrenceville. Then I continued talking to my friends and neighbors in the neighborhood as well as my council representatives and finding more and more people who felt the way I did, primarily in Lincoln Place – that was a good group that had been going on for awhile. And then I found throughout the city people were concerned and we began meeting on a semi-regular basis ... what happened with the Lawrenceville group is that very shortly after I learned about this and started reaching out to people in Lawrenceville, I got so involved with the greater citywide movement, that because we're all just doing this volunteer, my time was taken up completely by the citywide group. (Interview, January 18, 2011)

At the same time that residents were becoming aware of fracking, the Community Environmental Legal Defense Fund (CELDF) approached members of the Pittsburgh City Council with a draft of an ordinance to ban fracking, and Doug Shields subsequently sponsored the ordinance in the city council. Participants in the emerging anti-fracking movement testified at city council hearings and provided council members with information about the health and environmental risks of fracking, helping counter the industry's promise of jobs and economic benefits from fracking (Lord 2010).

With its November 3, 2010 event, Marcellus Protest demonstrated the value of a good tactic for mobilizing a movement. Organizers of the large demonstration in Pittsburgh intended the tactic to build the movement as well as to send a message to the industry and the state. David Meieran, a member of the initial organizing group, set up an excellent website, which was central to the movement's visibility. As described in Chapter 3, a series of open meetings were held to organize the protest, and working groups were formed around tasks such as fundraising, art, outreach, and logistics. A diverse collection of people came to numerous organizing meetings, including members of neighborhood groups such as the Lincoln Place Action Group, who were mostly new to environmentalism, members of established groups such as GASP and the Sierra Club, college students, radical environmentalists, and long-time progressive activists. Although most participants were white, they varied by social class, age, ideology, and movement experience. Young people, including members of the Shadbush collective, provided a great deal of energy to the organizing group. Long-time activists and experienced organizers lent their considerable skills to logistics and other matters. The numbers of people participating, with their mix of talents and perspectives, made for an exciting organizing experience.

Integrating newcomers is always a challenge for any movement, and active organizing of tactics and campaigns is very helpful in doing so. The lively working groups for the November 3 event provided easy ways to get involved, such as working on large banners, puppets, and signs with the Art group or sending out letters asking for donations and endorsements with the Outreach group. Moreover, an active and energetic group can motivate newcomers to return to the group, even if they are initially uncertain about participating. For example, Sara was a young woman who was just learning about the issue and started coming to meetings after the organizing was well underway. She initially found it somewhat difficult to fit in but was attracted by the group activity and eventually became highly involved in the movement. She recalled:

> I went to a couple of the [meetings] before November 3rd. I was kind of lost in them at that point because everyone knew about all this stuff. It wasn't really the ideal conditions for like a new person who was just coming to the issue to come in. Like everyone seemed to know stuff. People were mobilized around planning this thing, but I was still like able to get from what people were talking about that this was a serious issue and that it was something that I wanted to [get involved with] – I was motivated to self-educate and then I wanted to be involved with working for a ban. And at the time, there was just so much like positive energy around the movement, just like, oh, if we get this ban, it's going to mean so much to our region, like, it's going to like encourage other communities to take this step.... (Interview, June 27, 2012)

While many newcomers to movement groups often do not return after attending one meeting, Sara's experience shows how an exciting campaign can overcome feelings of awkwardness and motivate ongoing participation. Both the anticipation of success for the Pittsburgh ban and the demonstration tactic energized activists. The excitement of organizing the protest was contagious, and many individuals stepped up to provide resources and innovative ideas. For example, a member of the Art committee hosted weekly brunches at his home, at which donations were collected and participants talked about the issue and got to know one another. Through such interactions, a social movement community was forming.

Participation in both the planning and the protest itself created solidarity among activists, many of whom had not worked together previously. Ken Weir commented at a meeting following the protest (and in a later interview) that it was a "powerful" event and that walking over the bridge was an "amazing feeling." He said: "I was proud to be part of that group. I haven't felt that way since high school football! It was awe inspiring" (fieldnotes, November 12, 2010). Beyond inspiring participants, the protest brought people from the same geographic areas into contact with one another. Notably, several residents of Butler County who felt alone in opposing drilling in their conservative communities were introduced to one another at the protest event. They included Diane Sipe and others who joined a new Outreach committee formed by Marcellus Protest after the November 3 event and who eventually organized Marcellus Outreach Butler. MOB activists became highly important members of the area movement, who formed relationships with MP and Shadbush members and continued to support the movement in Pittsburgh and Southwestern Pennsylvania.

After its exciting start, Marcellus Protest and its allies experimented with various strategies aimed at building the movement and having an impact on shale gas drilling. The success of tactics generally depended on individuals taking the lead, and there was not always consensus on strategy. On November 4, 2010, the day after the big march, Loretta and Ken Weir organized a demonstration at the riverside to protest the illegal dumping of fracking water. When the action was proposed, it raised some concerns in MP about potential arrests and whether or not MP was prepared for this. These concerns were alleviated when the organizers offered assurances that they were not planning to get arrested; Ken said, "It will not be civil disobedience; if we're asked to move, we'll move" (fieldnotes, October 29, 2010). The action came off without incident and drew about thirty protesters who were enthusiastic about the demonstration and the media attention that it attracted. The organizers planned to continue with similar protests, but they never generated a sustained campaign around the issues of wastewater disposal and suspected illegal dumping, probably because it was very hard for a few people to organize such a campaign without a strong organizational infrastructure.

Marcellus Protest was fairly active for about three years, engaging in demonstrations in the Pittsburgh area and holding numerous educational events. Although there was not always complete agreement on tactics, individuals and groups could step up to organize protests, and many of the ongoing demonstrations were both fun and inspiring to participants. For example, when Shadbush led a protest with Occupy Pittsburgh and Marcellus Protest on the occasion of the DUG East (developing unconventional gas) industry conference in November 2011, it included flash mobs that danced in the lobbies of the hotels hosting the industry meetings. Loretta Weir recalled telling her incredulous adult daughter on the phone, "We're down at the subway station and we're practicing our flash mob" and how their nieces in California and Oklahoma were shocked to see her and Ken along with young people dancing at the flash mob on a video; "We had a blast," she said (interview, April 19, 2012). Loretta also led Christmas carol singing, with lyrics adapted to the fracking issue, for two years – another fun event. Enjoyable activities such as the flash mob and caroling provided opportunities for activists to expand their personal protest repertoires and to interact with other activists, attracting new participants and building solidarity in the movement. Nevertheless, such opportunities did not arise spontaneously; it was important that groups such as Shadbush or individuals like Loretta took the initiative to organize the events.

Although activists debated the extent to which they should become involved in providing services to affected communities as opposed to engaging in political tactics, direct service and public education were also mobilizing tactics for the movement. As described in Chapter 4, a number of activists told the stories of people harmed by fracking in the *Shalefield Stories* and other projects.

Members of MOB became very involved in helping families in the Woodlands area of Connoquenessing Township, Butler County after they lost clean water, and an emergency water bank was established through a local church (https://sites.google.com/site/waterforwoodlands). The Shadbush collective led several tours of affected communities, including the Woodlands. After getting to know residents of the Woodlands, Shadbush organized two large fundraisers to help the community. The collective also became close to Maggie Henry when they held an action camp on her family's farm and organized a subsequent action at the nearby drilling site threatening the organic farm.

As described in Chapter 3, the Shadbush collective was very active for several years and, along with MP, organized ongoing actions. Shadbush's collaborations with other groups reveal the importance of such alliances for maintaining movement mobilization and how participation in actions creates lasting ties in movement communities. In addition to memorable demonstrations such as the DUG East demonstration in 2011 and the pig action in 2013, Shadbush's participation was critical to the success of many actions done in cooperation with MP, including the 2012 demonstration focused on pipelines, the Tour de Frack event in 2012, the Earth Day demonstration at the DEP in 2013, and the demonstration against fracking in the county parks with Power Shift later in 2013. Relationships with MOB and other rural and suburban groups were very important, and activists from groups outside the city often traveled to Pittsburgh for MP meetings and events. One of Shadbush's goals was to introduce direct action to more activists, and their expertise was valuable to other groups. Shadbush and MOB developed a very close relationship after they worked together to stage a protest at state legislator Brian Ellis's office in 2012. That demonstration was a low-risk direct action at which experienced Shadbush activists showed MOB how to carry out such a demonstration, which made them excited to expand their tactical repertoire. Shadbush was also instrumental in initiating Protect Our Parks (POP), and its activists lent significant energy to some of POP's demonstrations and events, even though activists from Marcellus Protest, in coalition with the Sierra Club and others, ended up being the primary organizers of the campaign.

Protect Our Parks helped build the movement by educating the public about the threat to the county parks and by employing a series of tactics that allowed a wide circle of supporters to become involved in various ways. In the POP campaign, described in Chapter 3, activists in Pittsburgh worked tirelessly to bring the issue of fracking under the Allegheny County parks to public attention. POP supporters collected names on petitions against leasing the parks for fracking, testified at county council meetings for many months, met with council members, wrote letters to the editor, handed out flyers, sent educational materials to county councilors, held fundraisers, press conferences, and public meetings, collected signatures for a citizen's ordinance to place a moratorium on further fracking of county parks, and organized demonstrations. After early POP meetings that attracted more people, a core of only about ten people continued to regularly attend meetings, but the range of tactics employed by POP involved a much larger number of people in the movement community. As the slowdown in shale gas production due to a glut of natural gas and low prices put fracking in the parks on hold, the POP coalition stopped meeting after the summer of 2016, but remained prepared to jump into action when needed.

The issue of fracking in the parks generated community outrage and brought together grassroots environmental activists with people in affected communities, some of them new to environmentalism. As a result of POP's efforts, some residents who lived near Deer Lakes Park, the first to be leased, became involved in the struggle. After attending a community meeting with state senator Jim Ferlo, one resident began going door-to-door in her neighborhood collecting names on petitions against leasing the park. She and her husband attended some POP meetings, and she reported to the group that "people do care," and that the response she was getting "makes me go on" (fieldnotes, February 1, 2014). As in other movement activities, relationships between environmental activists and residents of affected areas were important, and participants in movement campaigns were energized by interactions with community members, who were becoming part of a growing environmental movement community. After the Allegheny County Council voted to lease Deer Lakes Park, POP continued to fight for the parks, and the resident who went door-to-door in her community told other participants in POP why she came to their meetings. She said that she and her husband go to Deer Lakes Park all the time; it is really a beautiful place. Even though it belongs to all of us, she said, nearby residents are the ones who benefit the most. She knows many people who live in the area care a lot about what happens to the park, but they haven't done much. But you all have, she said, and "I want to thank all of you for what you have done," how you stand up for the parks, and "that's why I continue to come" to the meetings. This was a heartfelt statement from a shy, soft spoken woman, and when she finished speaking, POP members all clapped and thanked her (fieldnotes, August 9, 2014). As this example shows, the power of the issue was enhanced by the relationships and emotion generated by the struggle.

Many tactics to oppose fracking, and the relationships they created, energized activists and made them feel inspired and empowered. Some efforts were also discouraging, such as POP's inability to convince Allegheny County Council to vote against leasing Deer Lakes Park for fracking. And regardless of such outcomes, it was not easy to keep the movement active and growing. For example, Loretta Weir found that MP's Action committee was able to put out a call on the email list for participants at various demonstrations and "you would get cooperation and then, eventually, you weren't getting anyone" (interview, April 19, 2012). Similarly, some general assemblies and educational forums attracted a fair number of participants, but others were disappointing with regard to attendance. While the issue of fracking and the tactics employed by activists helped draw people into the movement, organizational structures, resources, and leadership were also important in building the movement.

CONCLUSION

The above account of participation in the local environmental movement in Pittsburgh provides some insight into the processes involved in building a grassroots movement. Face-to-face interactions and events generated emotion, commitment, and identity. Mobilizing tactics, such as the demonstration in Pittsburgh on November 3, 2010 that helped build the anti-fracking movement, are critical in expanding the movement. Through their participation, supporters learn more about movement issues and about how to engage in particular forms of collective action. They also build relationships with other participants and become part of a grassroots movement community.

The history of the Pittsburgh movement demonstrates both the importance of ongoing strategies and tactics in building a movement and how difficult it is to keep organizing campaigns and actions. The movement expanded when it was able to devise tactics that provided a way for supporters to become involved in collective action. Many of these tactical opportunities came about in response to external threats, such as the activities of the shale gas industry and the decision of the Allegheny County executive to lease county parks for fracking. But the threats alone did not mobilize the movement; organizers came up with tactics that allowed for widespread participation. Even if particular tactics are not successful in persuading targets such as industry or government, participants often feel empowered by taking part in a public action such as a demonstration. Solidarity and collective identity are forged through participation in collective action, making it easier for organizers to mobilize subsequent participation. Media coverage, or social media attention, often makes supporters feel that they are having an impact by raising awareness and changing public opinion. Specific evidence of progress is also important; movement activists were encouraged by public opinion polls showing that growing numbers of people opposed fracking and by the Pennsylvania Supreme Court rulings on Act 13 that favored the movement. Yet there were also discouraging defeats, such as the inability to prevent the leasing of Deer Lakes Park, and it is hard to keep volunteer groups alive for many years.

To maintain a social movement for many years, it is necessary to continue to generate strategies and tactics and ongoing campaigns. This takes organization and the development of relationships in the social movement community and the creation of forums in which strategies can be devised. Chapter 6 lays out the chronology of the local movement during my study and examines efforts to create a movement infrastructure and community.

6

Creating Movement Infrastructure and Community

The environmental movement in Pittsburgh, like the national movement, has ebbed and flowed over time (Ogrodnik and Staggenborg 2016). Some local organizations, such as the Group against Smog and Pollution and the Allegheny Group of the Sierra Club, have survived since the 1970s, albeit with fluctuating levels of activity. Many other groups have come and gone, though they often have impacts beyond their lifetimes as active organizations, such as the creation of networks and changes in public discourse. The movement has experienced ups and downs as new issues motivate supporters and later recede in urgency. Individuals often become involved in social movements during periods of threat, when issues and organizing efforts are visible. As we saw in Chapter 5, some of the early opponents of fracking in Pittsburgh were newcomers to the local environmental movement who, as residents of city neighborhoods threatened by fracking, became aware of the issue. Movement organizers had to devise ongoing tactics and create organizational structures that would involve supporters and maintain the movement. This was difficult to keep up over time, and Marcellus Protest, Shadbush, and Protect Our Parks all eventually stopped meeting. Nevertheless, the movement continued.

This chapter takes up the question of how a movement changes and survives over time. Movements, by definition, endure beyond single episodes of collective action and consist of multiple campaigns and interactions with authorities and other targets (Tarrow 2011; Tilly and Wood 2013). To maintain themselves over time, movements need mobilizing structures that allow for the continuous marshaling of resources, particularly the recruitment and retention of participants, as well as ongoing strategies and tactics. This involves creating and maintaining not only organizations and leadership teams but also networks and relationships beyond organizational boundaries. People activated by movements often move from one organization to another and participate in activities organized by various groups and individuals. Alliances and networks of support are formed within movement communities, which include cultural as well as political groups and individuals as well as organizations. Focusing on the period of my study (from 2010 to 2017 with some observations extending to 2019), I look at how the diverse groups in my sample fared over time in the context of the larger environmental movement community.

CREATING A MOVEMENT INFRASTRUCTURE

Two of the organizations I studied, the Allegheny Group of the Sierra Club and Transition Pittsburgh, were connected to larger organizations and therefore had some guidelines for organizational structures – though these were very different for the two. In the case of Transition, the group tried to employ the flexible model devised by Transition founders based on "open space" organizing (Hopkins 2008, 2011) rather than creating more conventional organizational structures. The Sierra Club's Allegheny Group, in contrast, followed the lead of a long-standing national organization and state-level chapters, which provided more extensive guidance to local affiliates (e.g., Sierra Club 2016). Unlike Transition and the Sierra Club, the other groups in my sample were local ones that had to create organizational structures on their own. Marcellus Protest, Shadbush, and Protect Our Parks were all part of the movement against fracking in Pittsburgh that I was able to observe unfolding since 2010. In the following sections, I discuss the challenges faced by these different groups as they tried to create a movement infrastructure and some of the changes that occurred over time within movement organizations and in the movement community.

Transition Pittsburgh

Transition Pittsburgh attracted many potential members but struggled with how to create strategies and tactics that would organize its supporters and advance its goals. One difficulty was that the first Transition initiative in the United Kingdom was organized in the small town of Totnes, and organizers in Pittsburgh were uncertain about how to apply the model in their mid-sized city. Among the different types of Transition initiatives described in the Transition *Primer*, Pittsburgh organizers considered their group a citywide "hub" that would help form initiatives in local neighborhoods. There were numerous "Transition-y" projects in Pittsburgh (only a few of which were formally associated with the organization, but some of which were coordinated by Transition members), and these attracted participation by Transition members. There was no central direction or coordination of strategy, and Transition remained a loose model, which was nonetheless very attractive to participants seeking to create sustainable communities and local environmental projects. The international Transition model provided some guidance to the local group, and a national trainer from the Transition US organization traveled to

Pittsburgh to train activists on multiple occasions. Local activists also tried to create more structure by defining positions and associated responsibilities, but they never succeeded in building a leadership team, and a few individuals carried most of the organizational burden.

As described in Chapter 3, Transition Pittsburgh launched to great enthusiasm in 2010, but failed to remain continuously active and to spur the kind of neighborhood initiatives that some members hoped it would. The group experienced short-lived spurts of activity, but no ongoing campaigns. Because Transition was trying to create sustainable communities, rather than political change, members were involved in various projects such as community gardens rather than protest. The group did not have a good way to assess its strategies and therefore did not engage in much strategic adaptation. Participation in coalitions such as POP, organized by others, might have been one way for Transition to become more engaged with the broader environmental movement community, but the group generally did not participate in coalitions. This was partly because some members saw Transition as a nonpolitical group that should not become an ordinary movement organization and partly because it lacked the decision-making structure needed to actively participate in coalitions. Transition stopped meeting in 2013, though some individuals continued to organize a few events through Meetup until the end of 2016.

Allegheny Group of the Sierra Club

The Allegheny Group of the Sierra Club, in contrast to Transition, had much clearer guidelines about how to operate as a local group within a national structure, and the AG has maintained itself as a stable organization that engages in regular, ongoing activities for decades. The group was more active at some periods of its history than others, but always held monthly executive committee (ExCom) meetings. During much of the period of my study, two action teams were active and action nights were held regularly. Members of the action teams engaged in activities such as tabling and phone banking, and the AG endorsed political candidates, in cooperation with the Pennsylvania state chapter and the national Sierra Club, and worked on their campaigns. The group maintained some level of activity and participated actively in coalitions, but often struggled to recruit new activists and to launch its own campaigns.

While the Allegheny Group was a stable organization, it had its limitations.[1] During my observations, only a small number of AG members, perhaps thirty,

[1] As described in Chapter 2, Ganz and his colleagues conducted a study of state and local Sierra Club affiliates in cooperation with the national organization (Andrews et al. 2005, 2010). They found that the limitations of the AG are commonly experienced by other Sierra Club groups and chapters. In particular, many have difficulty activating more than 2 percent of their paper members, and most do little to develop leaders. Groups and chapters that are better at engaging members and developing leaders are able to exert more public influence, but, overall, the study

were actively involved on a regular basis, and the group had a difficult time recruiting members for its committees and ongoing work. One reason was that many people on the membership list were elderly, having been members of the Sierra Club for many years. Another factor was the decline of the AG's outings program, which was once a path toward involvement for many members. Recently, however, the state chapter hired a staff member who revived the program, organizing outings and hikes in Pittsburgh as well as surrounding areas. Some of the issue committees, such as the conservation committee, were very active in the past, and monthly program meetings attracted up to 100 people in the 1970s and up to 50 people in the 2000s. In more recent years, attendance at monthly program meetings dropped to about twenty, and the main organizer of the program meetings died. Consequently, the program meetings were discontinued until a new chair began trying to revive them in 2017 with limited success. The issue committees also became less active, consisting mostly of individual leaders without active committees or two or three persons who worked with other groups outside the Sierra Club. The action committees were small, with fewer than ten members each, and they tended to stick to a limited range of familiar strategies such as educational forums and tabling at public events. In 2015, the AG experimented with combining the action teams into a single Endangered Planet team to focus on several major campaigns (the Clean Rivers Campaign, Protect Our Parks, and endangered species). This proved unworkable, however, and the Endangered Species Action Team later reformed.

Different types of organizational structures create different advantages and challenges (see Table 2.1). The SC's federated structure presents both advantages and disadvantages in terms of strategic innovation. On the one hand, the federated structure creates some delays for decisions that must be approved, such as political endorsements, and sometimes adds to difficulties in breaking out of the standard tactical repertoire. National policies place some restrictions on group actions, such as a long-time ban on participation in civil disobedience (broken deliberately at the 2013 Keystone XL protest, where the national SC's executive director was arrested). On the other hand, the local group participates in many strategies devised by the national organization. For example, in February 2013, AG members organized buses to Washington, DC to join a nationally organized protest by the Sierra Club and other organizations against the Keystone XL pipeline. In September 2014, a number of AG members joined others from Pittsburgh in traveling to New York City for the People's Climate March. In April 2017, some members again went to Washington, DC to participate in the second People's Climate March to protest the Trump administration's policies on climate change, while others attended the sister march held in Pittsburgh. Allegheny Group members and staff began working in the Trump resistance

found that local Sierra Club groups and chapters could be much more effective. The national Sierra Club (2016) later focused more extensively on leadership in its organizing materials.

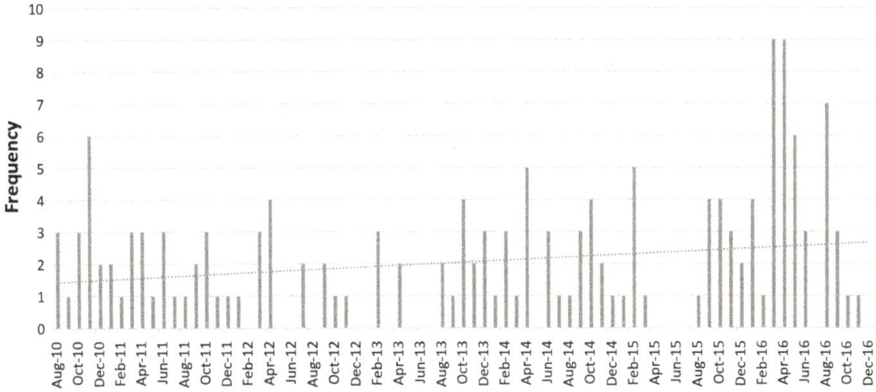

FIGURE 6.1 Marcellus Protest actions

movement after his election. Threats to the environment under Trump created opportunities for the AG to recruit new members, including younger ones. Staff member Eva Resnick-Day organized a large group that included many young people to work on the Sierra Club's campaign to get cities to commit to clean energy, and the AG welcomed several young members to its ExCom in 2017.

The Movement against Fracking

During the period of my observation, the movement against fracking was the most exciting development in the local environmental movement. Figure 6.1 uses posts on Marcellus Protest's main Google Group to chart actions that were either sponsored by the group or attended and endorsed by group members over time from August 2010 to December 2016. It shows some lulls in activity but also continuing actions over time. Individuals continued to learn of new actions through the MP network, including the Google Group and website, even after the Coordinating Committee ceased to meet in 2013. Figure 6.2 uses posts from Shadbush's email list to chart actions over time. There are some gaps in activity – what Shadbush members referred to as the "boom and bust" character of their activity – which often occurred after members wore themselves out in periods of intense activity and needed to pull back for a while. However, Shadbush did stay active for nearly four years. Figure 6.3 maps out POP actions using posts from the POP email list, showing continuous activities during its three-year lifetime, with peaks around major tactics and a decline after the group stopped meeting.[2] The email posts capture some of the fluidity

[2] Actions were counted if they met the following criteria: (1) there was an email or Google Group post explicitly stating an event was being sponsored by the group; *or* (2) there was a postevent email or Google Group post confirming at least one group member's attendance at the event, even if the event was not sponsored by the group; *or* (3) there was an email or Google Group post

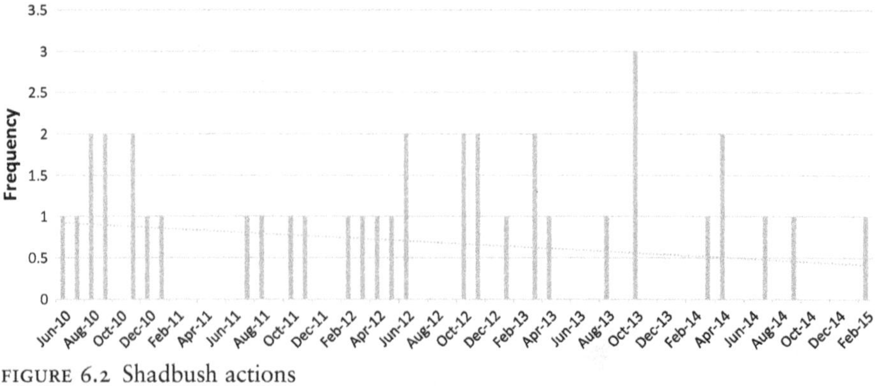

FIGURE 6.2 Shadbush actions

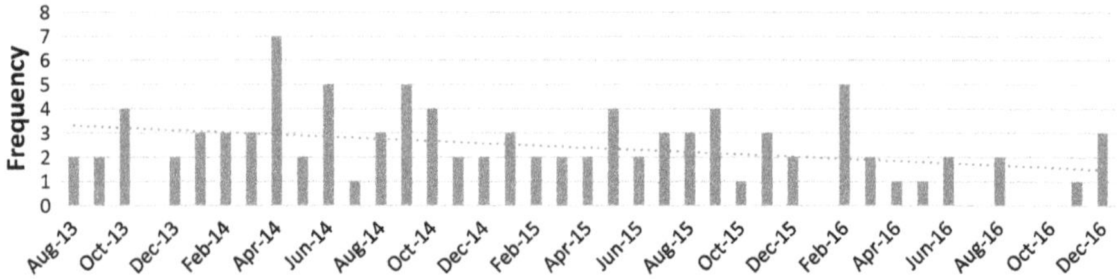

FIGURE 6.3 Protect Our Parks actions

of the movement and the importance of networks in the movement community; many individuals provide information about movement issues and events even when movement organizations cease to meet or become less active.

During the years of my study, the grassroots movement against fracking was fueled by the contributions of many individuals who were alarmed by the issue; the movement was a largely volunteer one that relied on the resources of its unpaid activists. Many people who had never before been environmental activists, as well as experienced activists, came forward to contribute their time, skills, and money to the movement. Mark Schneider of the Lincoln Place Action Group commented about the selflessness of people who were harmed by fracking and then spoke out to warn others, inspiring him and others to commit resources to the movement:

They need to have some kind of a civic award for their actions, their volunteer actions to help people ... They traveled everywhere for nothing, paying their own money to help

leading up to an event with at least one group member expressing firm plans to attend, even if the event was not sponsored by the group, and even if there was no email or post afterward to confirm attendance. Events were *not* counted as actions if the event was not explicitly sponsored by the group *and* there was no confirmation (pre- or postevent) of any group member's attendance at the event.

people, and that's incredible and that's what America's about. And that's why we did it. Ron Gulla, and Bob Donnan and Stephanie Hallowich actually inspired me to spend thousands of dollars of my own money to do this. (Interview, May 15, 2012)

Mark bought supplies such as a laminating machine, cameras, printer ink, and paper. He also removed the gas furnace from his home, even though it was more economical than electric heating. The Schneiders, the Weirs, and quite a few other activists in Marcellus Protest and the Sierra Club eventually installed solar panels on their homes, and more were looking into doing so. Many activists, including Ken and Loretta, spent their own money to travel to places such as Harrisburg and Philadelphia on multiple occasions to testify at hearings and participate in demonstrations. Mike Stout, a musician and long-time activist who co-owned a printing company, did numerous fundraisers for the movement, and he and his business partner also printed many materials at their own expense. Numerous other activists also contributed supplies and money as well as their time and skills to the movement.

Most financial and material contributions were small donations, but after the November 3, 2010 event, Marcellus Protest received an unsolicited donation of $10,000 from a foundation. This donation, together with a later donation of $5,000 from an anonymous individual, provided the group with financial resources for many activities. Marcellus Protest used these donations for its own work and to help smaller groups such as Marcellus Outreach Butler and Shadbush with particular projects. Thus, the anti-fracking movement had ample resources, but no institutionalized supply of money or staff. After the initial surge of movement activity, ongoing activism and solicitations were required to bring in more contributions. While some activists made significant personal sacrifices to contribute to the movement, there were no doubt others who lacked time or money to participate by, for example, funding their own travel. Marcellus Protest did support the participation of some such activists with small amounts of money, but relied heavily on people who could contribute resources, such as retired people with time, skills, and stable incomes.

A relatively small number of about twenty committed and talented individuals did much of the work needed to keep Marcellus Protest going – serving on the Coordinating Committee, organizing special events, attending hearings and demonstrations, putting out calls for participants for various events, developing new literature, speaking at public events, and keeping up the website, calendar, and Facebook page. After David Meieran could no longer keep up the website he created, John Detwiler, a retired engineer with computer expertise, stepped in as the webmaster, putting in many hours to maintain a critical tool for the movement. He also published an excellent newsletter from 2012 to 2016, cleverly billed as "All the Fracking News that Fits." Although a few issues were put out by guest editors, most were done by Detwiler, who researched and authored the bulk of the articles. The newsletter was distributed through the website and Google Groups, and everyone was encouraged to download, print, and distribute it, which some MP members did. Another volunteer, Gloria

Forouzan, kept up the Facebook page and website calendar, which covered events across the state, from 2010 to 2014. The MP website, Facebook page, newsletter, and calendar were important tools for the movement, providing a continuous online presence that allowed new people to find the movement even when there was not a lot of visible public action.

In fact, people from around the world contacted MP through the website, and these contacts helped connect Pittsburgh activists to others, both nationally and internationally. Soon after Pittsburgh passed its ban on fracking, activists in Mont-Saint-Hilaire – a city in Quebec, Canada – initiated contact with Pittsburgh activists that resulted in the two cities becoming "sister cities." After asking for MP's approval, John Detwiler took the initiative to work with Doug Shields on getting the city council to pass an official proclamation on December 6, 2011. John kept in touch with the Quebec activists, and he and his wife Claudia traveled to Montreal for a demonstration in 2012. In 2016, the contacts from Quebec asked MP to add its name in support of a protest against renewed threats from the shale gas industry. Delegations from three different countries – Denmark, Australia, and the Netherlands – also contacted MP asking for representatives to meet with them along with government and industry representatives when they visited Pennsylvania. A group of MP activists met with the Danish delegation, for example, and John Detwiler impressed them with a graphic presentation showing a map of Denmark superimposed over a map of Pennsylvania, which was about the same size. He then proceeded to show all the gas wells drilled in Pennsylvania each year and how it would look if they were being drilled in Denmark. Despite what has occurred in Pennsylvania, he told them, they may be able to stop fracking in Denmark – a warning that seemed to make a big impression. The Netherlands delegation came twice, and after the second visit, in July 2015, the Minister of Economic Affairs, who was part of the delegation, announced a five-year moratorium on shale gas drilling in the Netherlands.

Thus, MP had some important successes and gained visibility nationally and internationally, as well as in the state of Pennsylvania. A number of participants argued in interviews that MP helped raise consciousness and forced the media, industry, and government to pay attention to the opposition to fracking. Mel Packer said, "I think we've had a tremendous effect on just raising consciousness about environmentalism in general, but [also] about the whole shale process as being toxic" (interview, June 19, 2012). Briget Shields, who like Packer had been active in movements since the 1960s, commented:

> I think personally this group has accomplished more in a short amount of time than any other group I've ever been involved in. I'm always amazed at what Marcellus Protest has done, where we've come in such a short amount of time. It always gives me hope, you know. I think that it is depressing when you see that, you know, there's – I forget what the last number of leases that have just been recently signed. But then you see that the newspapers now each have their own Marcellus page and it's all over the internet. I think a lot of that is because of what we've done. You know, we've been very visible. We've

been very loud. We've been very focused. And I think that shows pretty much every time you see anything about that story, there's something of Marcellus Protest in it, whether it be a sign or one of our people. (Interview, April 3, 2012)

Public education was an ongoing goal, and activists regularly noted signs of the movement's progress, such as a Gallup poll showing a rise in opposition to fracking from 40 percent in 2015 to 51 percent in 2016 (www.gallup.com). In 2013, environmentalists celebrated when the Pennsylvania Supreme Court struck down major provisions of Act 13, a law that allowed gas companies to override local zoning laws, as unconstitutional (Hopey 2013a). In 2017, they celebrated again when the state Supreme Court issued a broad ruling based on the Environmental Rights Amendment, which declared that royalties from oil and gas leases on state forest lands had to be used for conservation purposes (Legere and Hopey 2017).

To remain active, movement organizations often devise mobilizing tactics to keep supporters involved as well as to create strategies (Reger and Staggenborg 2006:316). Beyond influencing the media and other targets, MP and other anti-fracking groups played a big role in educating and energizing their supporters. From 2010 to 2013, members of the Coordinating Committee worked hard to devise forums that brought people together to learn about the issues and debate strategies. The Shadbush collective contributed greatly to educating movement constituents and motivating them to take action by developing well-researched and informative materials, such as the *Shalefields Grassroots Reporter*. Marcellus Protest and other groups developed literature that was also distributed widely, and Aaron Booz, who was active in MP and POP as well as South Hills Area Against Dangerous Drilling (SHAADD), continued to update Shadbush's "Fracking 101" presentation and to make it available for group presentations. Other movement supporters also created educational materials and films such as *Triple Divide*, a 2013 documentary by Joshua Pribanic and Melissa Troutman about the impacts of fracking in Pennsylvania, which was shown widely by movement groups.

Despite all of these efforts and successes, it was difficult to create ongoing strategies and to continue to mobilize participants. Some strategic planning and actions came out of MP's general assemblies (GAs) (described in Chapter 7), but most did not result in ongoing campaigns. The CC planned general assemblies, often as educational events, but it was sometimes difficult to get very many people to come. When attendance was good and strategic ideas were generated, there was often little follow-up. While general assemblies were originally envisioned as the place where strategies would be formulated and decisions made, this structure proved inadequate because different people attended them, and there was little continuity from one assembly to the next. In 2013, the CC stopped having regular meetings because everyone was getting involved in POP, and there didn't seem to be a need for extra meetings. Nevertheless, activists stayed in touch, and former members of the CC who remained on the Google Group were occasionally asked to approve various appeals, such as requests for MP endorsements, in later years.

Many of the problems that MP encountered in trying to keep the organization going and in devising ongoing strategies and tactics had to do with leadership and structure. Marcellus Protest was largely decentralized with no strong leadership team to direct the organization. This structure allowed individuals to take initiative in organizing activities, writing blogs, and speaking to reporters, but there was not much organized development of strategy and consultation among activists. The problem was not a lack of talented and committed organizers. When asked in an interview if there was enough leadership in Marcellus Protest, Aaron commented: "There's a bunch of really good leaders. That's not the problem. Like, there are really good people. Somehow we just haven't been able to get good decisions out of good leaders. I don't know … Somehow we never really built a great group that did things out of those people" (interview, June 26, 2012). As this comment suggests, the group had difficulty making strategic decisions and devising new tactics.

As discussed in Chapter 2, Blee (2012) and Lichterman (2005) both found reflexivity and collective learning to be in short supply. Although MP tried at times to evaluate its strategies and structures, it had trouble making changes that would improve its strategic capacity. Marcellus Protest lacked the type of structure that might have facilitated the formation of a leadership team capable of providing strategic guidance and innovation to the group. Ganz (2000:1014–15) views leaders as "persons authorized to make strategic choices within an organization" and argues that effective strategies result from the interactions of members of a leadership team. He stresses that effective leadership teams need to be diverse, with ties to various constituencies, and that organizational structures that provide leadership teams with regular opportunities for discussion and that keep leaders accountable to one another and to their constituents are most likely to produce good strategies (Ganz 2000:1016). It is, however, extremely difficult for grassroots groups to create the type of organizational structure that can support effective leadership teams. Both organization and strategy are critical to ongoing mobilization.

Marcellus Protest was never able to create an effective and stable structure with its Coordinating Committee, general assembly, and working groups. This was due in part to the diversity of views among participants about the structure required and the lack of experience on the part of many volunteers with creating such an organization. There was some ambivalence about the value of trying to make MP into an organization at all, particularly after trying to do so for a few years, and there were disagreements about the usefulness of holding regular meetings. The role of the CC was the subject of some mild disagreement at the start, which later resurfaced. Initially, the committee was seen by some as an administrative working group that would attend to running the organization, just as the logistics group had operated for the November 3, 2010 demonstration. Others talked at the start about a steering committee, possibly with elected members, to provide leadership to the organization, but participants who favored more "horizontal" structures rejected this idea. At a GA in early 2011, one activist responded to a suggestion that the group might need a more formal "board" with the comment that "a certain proportion of this group will not stand for a steering committee or a board or any language that suggests this type of thing, which is why we agreed on the term coordinating committee. We want to avoid the idea of a group with power, but instead want one that coordinates activities." Another activist replied: "We are split on this question. Some people want more organization" (fieldnotes, January 8, 2011).

There was no resolution of this dispute, and throughout the time in which it was active, the CC struggled to structure itself and to build MP as an effective organization. While it was supposed to consist of representatives of working groups and member organizations such as Shadbush, it was difficult to get all of these groups to send a representative to every meeting. Despite having committed members who took turns serving on the committee until the fall of 2013, CC members did not interact regularly outside of meetings, and most did not feel they had the authority to generate strategy. They did make some decisions for the organization, such as authorizing expenditures, usually small amounts of money. But given the lack of shared understanding of the role of leadership in the group, MP was unable to create a leadership team or a way of developing new leadership. Instead, the group ran on individual enterprise, depending on those willing to step up and organize events. The working groups, which included Outreach, Action, Internet, Strategic Messaging, and a few that never quite got off the ground, such as a Media group, were intended to drive the organization, but they met with mixed success. Some met only sporadically and were short-lived, while a few met regularly until they fulfilled their missions.

The most vital working group to form out of the November 12, 2010 organizing meeting following the November 3 demonstration was a reconstituted Outreach committee, with a number of new members from rural areas of Southwestern Pennsylvania, as well as from Pittsburgh and its suburbs. The Outreach committee met regularly for a little over a year, often attracting over a dozen participants from a variety of different regions and groups. As a result of the work of the Outreach committee, several rural or suburban groups formed outside of Pittsburgh in areas directly affected by shale gas drilling. The Action working group organized a number of small-scale demonstrations in response to industry events and other developments. However, the Action committee did not continue to meet for very long, in part because participants disagreed about the need to meet. Some organizers felt that meetings were a waste of time and that calls for action should simply be put out on the email list when needed. This worked for a while, when individuals organized targeted demonstrations, but did not produce any long-term strategy or firm commitments to the organization. Eventually, fewer people showed up for the one-off demonstrations.

While some working groups functioned for a time and then fizzled out, those that had clear missions, and those in which members forged strong relationships with one another, appeared to enjoy the most success. Participants in the Outreach group, during the year when it met, enjoyed one another's company and felt they were achieving goals. When Outreach disbanded, most felt it had fulfilled its purpose of raising consciousness and generating a network of groups and individuals inside and outside of Pittsburgh. Ties created by the working group, such as those between Shadbush members and Butler County activists who formed MOB, endured after the Outreach committee ceased meeting. The Strategic Messaging committee, described in Chapter 4, was another successful working group. The working group not only had a specific task but consisted of a congenial group of people who enjoyed meeting together, discussing works on strategic messaging, and thinking about how to be effective. Once its mission was completed, after about seven months of meetings, the working group disbanded. In general, it seemed easier to create groups as they were needed for particular tasks, rather than to form standing committees that might generate new tasks or actions. As in the case of the Strategic Messaging committee, MP activists followed developments in the battle over fracking in Pennsylvania and other states and often responded to industry strategies and governmental decisions.

To some extent, Marcellus Protest was a victim of its own success; once the ban on fracking was passed in Pittsburgh, there was no longer an obvious strategy to pursue within the city. Neighborhood groups that formed in Pittsburgh generally did not take a NIMBY (Not in My Backyard) position, but the ban did reduce the immediate threat that residents felt when they thought fracking might take place in the city. After the MP Outreach committee spawned several suburban and rural groups, many of their members became less active in MP before the Outreach committee dissolved. Following the burst of activity leading up to the November 3, 2010 demonstration, some participants were exhausted and ceased their participation in the movement. Some remained in the movement but chose to put their energies into more ideologically coherent groups. For example, Patrick, an important organizer of the November 3 demonstration, announced afterward that he was stepping back from MP to put his energies into the Shadbush collective, which was more in line with his radical leanings. Marcellus Protest was more ideologically diverse, with participants coming together around their concerns about fracking and environmental issues but disagreeing on matters like the value of electoral politics. Nevertheless, there were plenty of frames that unified the group, such as the importance of community rights and the need to preserve the environment for future generations.

While there were external constraints on strategies, MP continually tried new ideas and coordinated with other groups, notably Shadbush. Many ideas were partially explored, or generated a few actions, but failed to produce ongoing campaigns. One reason for this was disagreement about targets, particularly when the issue was complicated. For example, some members of MP were alarmed that farmers in Pennsylvania were selling their drilling rights without understanding how fracking might damage their water, animals, and crops. Some participants suggested boycotts of farms that allowed fracking, but others wanted to avoid blaming farmers, who had a difficult enough time earning a livelihood and who might not have understood the implications of leasing their land for fracking. MOB, the Butler County group that emerged from the MP Outreach group, organized a demonstration at a drilling site on a dairy farm in the summer of 2011, but was careful to blame the gas industry rather than the farm in its messaging. MOB did not want to endanger its own reputation in rural areas and would have liked MP to lead a campaign around farming, but MP lacked the capacity to do so. While there was ongoing talk in MP of a "frack-free farming campaign" to support farms that did not allow fracking and the CC at one point agreed to support such a campaign (fieldnotes, September 1, 2011), it never got off the ground. Beyond the minor disagreements over targets, this was mainly due to a lack of entrepreneurs available to spearhead such a campaign.

The Role of Alliances

Alliances among movement organizations were an important source of ongoing mobilization, though there were also limitations to alliances in the anti-fracking movement. Cooperation between Shadbush and MP was particularly important in carrying out actions, and several members of Shadbush played a critical role in supporting MP for the first two years, serving on the CC and participating on the Outreach committee. Most Shadbush members came to MP events and demonstrations, but not all members of the collective were enthusiastic about MP, given its lack of cohesive membership and ideology. As MP began to decline, Shadbush was increasingly reluctant to spend time helping to prop up MP. While Shadbush was energetic, it was small in size and, therefore, limited in its capacity, making members cautious about overcommitments of their resources.

In addition to its collaborations with Shadbush, MP also teamed up with other groups. For example, the Allegheny Group of the Sierra Club had an annual tradition of marching in Pittsburgh's St. Patrick's Day parade, usually with a theme such as clean energy. In 2011, the AG's Global Warming Action Team (GWAT) decided to make shale gas drilling the theme for that year and invited MP and other groups such as the Pittsburgh Student Environmental Coalition (PSEC) to join them in planning the event. Several planning meetings were held, and the outside participants, mostly from MP, more than doubled the size of GWAT members and energized the group. One MP participant argued that the parade contingent should be "loud, large, and fun" (fieldnotes, January 25, 2011), and it turned out to be highly successful in this regard, creating a colorful, large presence with a snake puppet, signs, and a marching band. However, one of the signs used to depict fracking had a skull and crossbones, alarming parade organizers who wanted the event to be "family friendly." Consequently, the parade organizers rescinded the AG's longstanding invitation to the parade for the following year. This led the AG's chair to meet with the parade organizers, who agreed to reinstate the parade invitation the following year with the understanding that the Club should be less controversial in its participation.

The AG was somewhat divided in its reactions to these events, with some members expressing a certain amount of glee about the demonstration that got them "kicked out" of the parade, while others were concerned about the group's reputation. Even though MP was considered more radical than the Sierra Club, it was, in fact, a Sierra Club artist who created the offending image, though the attention-getting band and protest-like participation were no doubt part of the concern for parade organizers. In 2014, there was some talk of joining with POP in making protection of the county parks a theme, which would have made a splash again, but AG leaders decided to focus instead on the Clean Rivers Campaign to find green solutions to sewage runoff into the rivers – a completely reasonable, if less controversial, choice. In the following year, however, the Club decided to suspend participation in the St. Patrick's Day parade because it was a lot of work without clear payoffs. Instead, they decided to participate in the gay pride parade, which would help create ties to another movement constituency. The AG was warmly welcomed by gay pride organizers, who were not likely to share the concerns of the St. Patrick's Day organizers about the nature of the Club's participation. After the first year of participating in the gay pride parade, the Sierra Club joined an alternative gay pride march in protest of the original march organization's decision to associate with EQT, the Pittsburgh-based natural gas company engaged in extensive fracking.

The Protect Our Parks campaign was another opportunity for coalition work that expanded the movement. The Allegheny Group is very open to participating in alliances, and several of its members were actively involved in the coalition to oppose plans by the Allegheny County executive to sell rights to drill under the county parks for shale gas. The coalition brought together leaders from a number of environmental organizations as well as newcomers, and Sierra Club members participating in POP kept the AG, which signed on to the coalition and donated money to it, informed of its efforts. Participation by a variety of groups, including MP, Shadbush, SHAADD, the Sierra Club, and GASP, among others, was important to POP's ability to get many people to testify before the Allegheny County Council for a number of months, and to bring out participants for the campaign's many other activities. Eventually, however, the core group of about ten regulars at meetings came to seem like a collection of individuals rather than a true coalition of organizations. In fact, participants decided that the groups that had originally signed on to a letter as supporters of POP should be asked if they still wanted to be considered organizational members and if they considered the participants from their groups to be official representatives. While not all of the coalition "members" responded, some were apparently comfortable with the arrangement; in any case, the discussion suggests a decline in the power of the "coalition" over time (fieldnotes, August 1, 2015). Coalitions of organizations are certainly important to collective action, but individual participants willing to put time into attending meetings and devising strategies and tactics are the backbone of a grassroots movement.

MOVEMENT COMMUNITY NETWORKS AND RELATIONSHIPS

Relationships among movement participants, both within organizations such as the Sierra Club and between groups such as Shadbush and MOB, are critical to ongoing involvement in grassroots environmentalism. Even when particular organizations stop meeting, relationships and networks in the larger movement community keep the movement alive. And despite the importance of organizational structures, the initiatives of individuals are often critical to the success of activities and campaigns that maintain the movement. Individual initiative is needed to carry out actions, and movements provide individuals with ways to use their skills and creativity. Individual efforts, however, need to be supported by other participants in the movement community in order for collective action to occur. Collective actions and campaigns are essential for building the movement as well as having an impact on targets, and they often involve coalition work among groups in the movement community. Individuals who do not regularly attend organizational meetings often come out for demonstrations and events.

During the several years in which the local movement against fracking was quite active, ongoing actions and campaigns helped build relationships within the movement community. In addition to the ties forged among groups through cooperation in collective action, individuals from different groups met when they participated in actions, campaigns, and coalitions. Often, people who did not know one another previously, and who may have worked in groups with different styles, got to know one another as a result of cooperative actions. For example, Briget Shields said she knew very few of the other activists prior to her involvement in the anti-fracking movement; this changed dramatically, she said, "And it's like we've all turned into a big family now" (interview, April 3, 2012). Few people in Marcellus Protest knew very many activists in the Sierra Club until the AG invited MP and other groups to raise public awareness of fracking by participating together in the St. Patrick's Day parade in 2011. The AG later provided opportunities for informal interaction by inviting MP members and others to their picnic and holiday parties.

Participation in collective action is one important way in which ties are created among diverse movement constituents and movement communities are built. An example of this occurred in 2013 when MP and Shadbush organized the Earth Day protest at the Department of Environmental Protection (DEP) office in Pittsburgh. While Shadbush had a reputation as a radical environmental group, Sierra Club activists were among the participants, as were representatives of other established environmental groups. Tom Hoffman, who was then the Western Pennsylvania director of Clean Water Action

(CWA), met Patrick from Shadbush at the event and told him that he was very impressed with the protest. Tom expressed interest in what Shadbush was doing, and the two met for lunch on another day to talk more and get to know one another (interview, June 17, 2013). Later, when Shadbush was discussing who to invite to a planning session for launching a new campaign, Patrick mentioned the encounter and asked if it was okay to invite Tom, who is very knowledgeable about water issues, to the planning session. Other Shadbush members agreed that this was a good idea, even though they had not previously worked with Tom or the more mainstream CWA (fieldnotes, May 4, 2013).

The coalition campaign that came out of that effort, Protect Our Parks, introduced people from different groups to one another and allowed participants to learn more about organizational members. For example, a young woman in POP thanked CWA for its lawsuit challenging the dumping of wastewater from fracking into Pennsylvania rivers (fieldnotes, November 1, 2013). People from different organizations, who didn't know one another previously, volunteered to work together on tasks such as writing a press release or creating literature. At the first open meeting of the new POP coalition, which was attended by a diverse group of people, I observed participants exchanging contact information with one another and offering to put one another in touch with people they knew (fieldnotes, August 3, 2013). At the next meeting, Mel, from Marcellus Protest, mentioned that Matt, from the Sierra Club, had sent him a great email about "guerilla actions," and they were going to sit down and talk about it after the meeting (fieldnotes, August 17, 2013). Individuals who did not regularly interact with one another were forming relationships through the campaign. Moreover, individuals who had been active in the movement but were not regularly attending meetings came to some of the early POP meetings and got involved in the campaign by testifying before the county council, collecting signatures on petitions, and other tactics.

Actions and campaigns provide ways for individuals to become involved in movements, enlarging the core of committed participants in the movement community (Downton and Wehr 1991). While collective campaigns are needed to build movements, individual initiative is necessary to maintain movement organizations and launch actions. Although Marcellus Protest had difficulty creating a workable organizational structure, individuals often stepped up to do the organizing and logistical work needed for actions. For example, Loretta Weir put out calls on the email list for people to participate in various demonstrations. John Detwiler responded to international requests made through the website and organized several meetings with international visitors to the Pittsburgh area. Mel Packer took on a lot of logistical work, and he often called people to ask them to participate in actions, such as the sit-in at the county executive's office that took place in October 2013 as part of the Protect Our Parks campaign. In all of these examples, individuals took initiatives on behalf of the group, but they were very much a part of the group and its actions. In other cases, and in other organizations, some individuals act as "lone wolves" who prefer to take actions on their own, a style of organizing that does not help to engage others in the movement (Han 2014).

All movement organizations rely on individuals to take responsibility for carrying out the tasks necessary to collective action. Groups vary in the extent to which such individuals are integrated into the group and accountable to it and in the extent to which activities are directed by the group. When individuals who take initiative also participate in the organization in other ways, they typically have some accountability to the group and are trusted by other members. In Shadbush, individuals were tightly integrated into the collective; there were really no lone wolves, though there were people who took more responsibility than others at times. Protect Our Parks was similar by its second year when a core group met together frequently and decided on actions collectively. In MP, the structure was much looser, and some individuals created projects that were independent of the group. In the Sierra Club, even though the AG had a well-defined structure, some issue chairs were lone wolves who worked independently. In the case of Transition, the organization was always a fluid one that loosely connected different environmental projects in the community (Staggenborg and Ogrodnik 2015).

Although formal organizational structures were difficult to maintain in the movement against fracking, more informal networks and relationships between individuals and organizations have endured. Staff members of groups such as the Sierra Club are typically very well connected to staff of other organizations, and they often strengthen the movement community by connecting activists from different groups to one another. For example, Randy Francisco told the ExCom about events he was organizing with other groups, such as the Moms Clean Air Force (fieldnotes, August 13, 2012), and he often provided the names of contact persons to AG members such as a law student who wanted to get in touch with Fair Shake Environmental Legal Services (fieldnotes, January 9, 2017). Campaigns and actions may also create lasting networks and relationships among groups (Wood et al. 2017). Importantly, ties among activists in the local environmental movement have been formed and strengthened as a result of the mobilization against fracking. For example, after Shadbush disbanded, contacts among members persisted. Robin told me that another Shadbush member had helped her get a job in a community organization and noted:

> One of the cool things I think has come out of Shadbush is those of us who were core members, like we still keep in contact somewhat, and even if we don't talk a whole lot I think there's still like a network there.... So, you know, we're all doing very different things in the activism world, but like through each other we know more about what's going on... (Interview, April 9, 2015)

Relationships between individual Shadbush members and groups such as MOB also endured, and activists continued to learn from one another. Jessica reported at a Three Rivers Rising Tide meeting that MOB had invited her to facilitate their strategy retreat and that she was "inspired by their framework" of working with community groups, mentoring people, and pursuing larger goals of banning fracking and promoting renewable energy (fieldnotes, February 28, 2015).

Networks and projects pursued by Transition Pittsburgh also persisted when the organization declined. After Transition stopped meeting regularly, its members continued to try to build community through events such as skill sharing, gift circles, and potlucks and a wide variety of sustainability projects. I also spotted Transition members as well as those of other environmental groups at meetings of Pittsburgh 350.org, which gained importance in the local movement community as activists tried to figure out how to battle climate change. One of the founders of Transition Pittsburgh, Elisa Beck, also continued to work through her neighborhood group, Sustainable Monroeville, on issues that in 2017 included community restrictions on seismic testing by shale gas companies. Maren Cooke, a Transition member who has also been active in GASP and other environmental groups, began holding educational and social "Sustainability Salons" in her solar-powered home in 2012 after talking with another Transition member about the need to promote interaction and education in the environmental community (interview, April 15, 2014). Although the Sustainability Salons were never a Transition project, they reflect some of the spirit of Transition with its desire to promote community, and they have attracted Transition members as well as others in the environmental community. The salons, which consist of potluck dinners and speakers or films about environmental issues, are attended by people from various environmental groups in Pittsburgh, including Transition, the Sierra Club, GASP, and others. Speakers from different groups have talked about projects and coalitions such as the Clean Rivers Campaign, and individuals who attend regularly get to know one another. The salons are still held monthly to date, providing a forum for regular interactions and relationship building in the movement community, outside of any single organization. They frequently draw more than fifty people and allow participants to network and find ways to become involved, either through individual actions such as installation of solar panels or through participation in environmental collective actions (see http://marenslist.blogspot.com/).

While MP and other groups focused on the issue of fracking, activists had much broader environmental and progressive concerns, and they supported one another in a range of causes. For example, Ted Popovich was a long-time member of GASP who also became active in MP and POP and in a group called Allegheny County Clean Air Now, which opposed the Shenango Coke Works plant that was eventually closed. He updated POP members regularly on this activity, and some of them joined him in testifying and targeting organizations such as the Allegheny County Health Department with regard to the health impacts of fracking and other pollution. Later, Popovich and others, including Sierra Club members, tried unsuccessfully to block Shell's efforts to build an ethane "cracker" plant in Beaver County. Local activists continued to protest against Shell and its plans to build additional plants in the area, holding a "Stop the Madness" rally and march against the toxic gas and plastic infrastructure in Pittsburgh on June 20, 2017. The protest, which was organized by MP and other groups, was promoted on the MP Google Group and on other email lists, including that of the Sierra Club. The protest called attention to the air and water pollution expected from Shell's planned ethane-cracking plant on the occasion of the National Petrochemical Construction conference and the Developing Unconventional Gas conference (Hopey 2017b). Groups of residents, such as the Beaver County Marcellus Shale Awareness Committee, also continued to protest the Shell plant and pipelines as they worked to protect the area water supply (Fawcett 2017). When President Donald Trump visited the Shell plant in August 2019, some 200 people protested, including members of MP, POP, and other environmental groups (Routh 2019a). On October 23, 2019, when Trump returned to the area on the occasion of a shale gas industry conference in Pittsburgh, he was again met with protest, by both environmentalists and Jewish activists protesting white nationalism and connecting Trump's rhetoric to the murders of Jews at the Tree of Life synagogue in Pittsburgh one year earlier (Bradbury 2019; Routh 2019b).

Thus, even when there was an ebb in anti-fracking activity in Pittsburgh, the flow of the larger area environmental movement continued, and activists who worked together in opposition to fracking maintained ties with one another. After groups such as Marcellus Protest and POP stopped meeting, participants stayed in contact through email lists and became involved in new actions from time to time. For example, Mel Packer, John Detwiler, and others became involved in 2018 in advising residents in the nearby town of Braddock, which was facing renewed threats of fracking after having previously stopped efforts to frack at a golf course there (email communication, February 7, 2018). In Fox Chapel, Dianne Peterson, a resident who had previously been active in MP and POP, organized an educational event and ongoing campaign in 2018 against fracking in the Pittsburgh suburb. In April 2018, numerous local activists attended an industry event to demonstrate their opposition to fracking and fossil fuels. In 2019, a Petrochemical Action Team was created by 350.org Pittsburgh to continue organizing around the issue of cracker plants, attracting members from MP, POP, and other groups. The movement community was clearly alive and well.

The actions and campaigns of the anti-fracking movement created both models of protest and connections among activists that provide opportunities for new and revived organizational activity in the movement. In 2018, a group of citizens founded an Allegheny County chapter of the Izaak Walton League as an organization that could potentially reach rural as well as urban residents of the county to address issues such as water quality and conservation. Mike Stout, an activist who had previously worked with MP and POP, became president of the chapter and one of the leaders of an effort to create a new countywide coalition of environmental, health, and labor organizations.

As Mike explained at a June 2019 meeting of the Sierra Club at which he invited the AG to become a coalition member, he was interested in targeting the county council by taking advantage of the recent election of some progressive councilors and working to further change the composition of the council. Protect Our Parks provided a model of a campaign that mobilized many citizens with tactics for influencing the county council. Although POP failed to prevent fracking in Deer Lakes Park, changes in the county council – and the potential for further changes – created new opportunities for successful actions targeted at the council (fieldnotes, June 10, 2019). One of the people in touch with Mike Stout on this initiative was John Detwiler, another activist in MP and POP, who called a meeting of the core members of MP and POP in the summer of 2019 after attending a meeting of the new Allegheny County Environmental Health Coalition. He described how Mike credited POP with going to county council meetings and creating public awareness of the issue of fracking in the parks, which made John think more about organizing around health and the environment (fieldnotes, June 22, 2019). At this meeting, attended by ten people, the group decided to revive itself as Marcellus Protest, which had a broader mission than POP.

CONCLUSION

This chapter shows how grassroots environmentalism in Pittsburgh flourished, survived, and partially declined over a period of seven or more years. The movement expanded during this period as the issue of fracking stoked passions and organizers created tactics to build the movement. Within organizations and the movement community, individuals developed relationships with one another that sustained their participation. Grassroots environmental groups organized many innovative and memorable activities, which enlarged and maintained the movement. Coalitions and campaigns expanded the movement and strengthened ties in the movement community. However, it is difficult to continually devise tactics that will bring out supporters, and grassroots organizations frequently decline after a few years. In my sample, only the well-established Sierra Club continued to meet regularly after POP stopped meeting in 2016.

The internal structures, resources, and cultures of movement communities and organizations, as well as external events and political opportunities, are important to the ability of the local movement to remain mobilized over time. Tactical innovation often occurs through the interactions of movement activists and their opponents (McAdam 1983). It also happens when movement actors interact with one another, either through alliances and networks in social movement communities or within social movement organizations. In the Pittsburgh anti-fracking movement, the industry and its government allies often provided targets for activists, and movement groups frequently worked with one another to stage protests and carry out campaigns aimed at these targets.

When threats became less visible, it was more difficult to continue to find compelling targets. In the face of either success, such as the Pittsburgh ban on fracking, or failure, such as the inability of POP to prevent drilling at Deer Lakes Park, it was hard to keep up movement activity. Nevertheless, the movement continued in various ways despite internal and external obstacles because both environmental concerns and the movement community persisted.

To endure, grassroots movements need to create movement infrastructure – a community of activists, organizations, and other mobilizing structures. Mobilizing structures include relationships and networks as well as organizations, which are important to the ability to expand and maintain movements. Depending on how movement organizations are structured, they may bring together leaders and other participants to regularly devise new strategies and tactics. Relationships within the movement community help create alliances that reinvigorate local movements and organizational teams sometimes take shape. MOB and Shadbush, for example, formed a team that engaged in several innovative actions. The internal structures of organizations also affect their ability to develop commitment and to creatively design and organize protests. While different movement organizations face many of the same problems, such as difficulties in recruitment, their structures and cultures facilitate some types of organizing and limit others. For example, the Sierra Club's Allegheny Group and Shadbush both had difficulties expanding their active cores, but for different reasons. The AG did not demand that members engage highly in the organization and provided instead minor ways for them to participate. Intended to facilitate low-engagement participation, this strategy may fail to produce enough committed members (Han 2014). Shadbush asked much more of its members and invested in the development of their skills, which generated commitment, but the group suffered from a lack of inclusiveness in its zeal to find like-minded participants. More diverse groups, such as Marcellus Protest, may have trouble agreeing on structures and frames, but even ideologically cohesive groups such as Shadbush have unanticipated problems in organizing and creating an inclusive collective identity. The following two chapters examine these dynamics in the meetings and forums of movement organizations.

7

Social Movement Meetings

Collective action is critical to social movements; tactics are necessary both to mobilize participants and to impact targets. Through tactics such as demonstrations, movements involve supporters and make their concerns known to bystanders and authorities, raising public awareness and influencing public policy. Some collective action occurs relatively spontaneously (Snow and Moss 2014), and social media have increased the ability to organize protests quickly when emotions are running high, as in the case of the Trump resistance movement (Meyer and Tarrow 2018). But ongoing action typically involves ongoing talk that occurs in face-to-face settings as well as online. Many important processes occur within social movement organizations (SMOs), and organizational dynamics are essential to the growth and maintenance of movements (Blee 2012; Ganz 2000; Han 2014; Lichterman 1996, 2005; Polletta 2002). Activists attend meetings, commit resources, choose leaders (or avoid doing so), plan actions, and recruit and integrate newcomers into their organizations. Yet, we still do not know a lot about how the meetings of different types of organizations facilitate or impede these activities (Baggetta et al. 2013:564); my contribution in this chapter is to show what happens at movement meetings and how they affect mobilization and strategic decision-making.

While actions and campaigns are essential for movement building, mobilizing structures also affect ongoing collective action. Movement organizations develop internal structures and cultures, either intentionally or by default, which affect their ability to maintain themselves and to engage in activities that advance their goals. Many challenges are common to grassroots organizations, but they take different forms depending on organizational structure, culture, and ideology. Internal organizational dynamics, as well as interorganizational relations, strongly affect strategic capacity and tactical choices as well as the mobilization of support. SMO structures differ with regard to the number and types of meetings that are held, their decision-making processes and mechanisms for implementing decisions, the types of organizational committees created, their means of securing resources, and the offices or roles available to leaders and activists. Organizational cultures vary in terms of how much participation is expected, the nature of the discussions, the formality of rules and procedures, the types of emotions that are expressed, the rituals that are followed, and the relationships that develop among participants.

Chapter 3 provided some general description of the organizational structures of my sample SMOs and narratives of their activities. The following three chapters then examined processes of mobilization in the grassroots environmental movement: issues and framing (Chapter 4), how individuals become movement participants (Chapter 5), and the creation and maintenance of movement infrastructure (Chapter 6). The present chapter goes inside the meetings of groups in my study to look closely at their internal procedures and cultures. Its detailed description reflects the extensive and complex discussions of activists, revealing the difficulties they face in creating structures, engaging activists, and making strategic decisions. This discussion is followed in Chapter 8 by an examination of the process of devising strategy within organizational meetings and at special forums organized for the purpose of generating strategy and improving strategic decision-making. As we will see, the sample organizations adopted different approaches to core organizational concerns and experienced related advantages and disadvantages. My observations of meetings and strategy forums show that there are no perfect models of grassroots organization and strategic decision-making processes, but organizational choices do influence strategic capacity.

Social movement meetings are typically central to organizational planning and decision-making. Although social media and online organizing are increasingly important to "connective action" (Bennett and Segerberg 2013), they usually supplement rather than replace face-to-face organizing in grassroots movements. Movement organizations hold various types of meetings, which present different challenges. These include relatively large meetings, sometimes called general assemblies (GAs), that are typically open to new participants and that may have either open agendas, as in the case of Transition's open space meetings, or specific tasks and goals, as in the case of the early Marcellus Protest (MP) meetings held to plan the November 3, 2010 protest. At the other end of the size spectrum are small meetings of collectives, such as Shadbush, or committees, such as the MP Coordinating Committee (CC) or the Sierra Club (SC) Executive Committee (ExCom), which vary in size and inclusiveness but are typically open only to members. Rituals and decision-making rules also vary, with some groups such as the Sierra Club using parliamentary procedure and others using majority voting or consensus decision-making. Meetings may be run more or less informally, and they may use conventions such as "keeping stack" (a list of people who have indicated they wish to speak) and rules such as quorums. Meetings also vary a great deal in length, with some adhering to a strict time length (sometimes dictated by how long a room is available) while others are open-ended. Some meetings are held in members' homes, some in community spaces such as churches or public libraries, and some in office spaces such as the Sierra Club's field office. Some meetings are easy to participate in, whereas others seem more exclusive. And some meetings are interesting or even exciting, while others are boring or tedious.

What makes for a good meeting? When asked this question in an interview, Jessica, a Shadbush member who also attended many Marcellus Protest meetings, offered:

> A good meeting is if you get things accomplished and if the people who participate in the meeting feel like they're reasonably engaged and feel like the process is comfortable enough for them – maybe not every minute, but like they don't feel alienated, they don't feel disrespected, and they feel like it's possible for them to engage as they want to. Then, you know, maybe a great meeting is when it would actually build a sense of teamwork and solidarity and people feel like they achieve a lot. (Interview, May 4, 2015)

Movement organizations try to hold meetings that cover topics that need to be discussed, make necessary decisions, allow participants to contribute ideas, and leave everyone feeling that the time was well spent. Beyond these basics, good meetings can foster relationships among participants, create collective identity and commitment to the group, teach attendees new skills, provide motivating information, and generate fresh ideas. Difficult meetings, on the other hand, might be overly long, boring, badly organized, poorly attended, unproductive, repetitious, or conflict-ridden. Of course, views on what constitutes a "good" or "bad" meeting vary somewhat with the culture and ideology of the group. Collectives such as Shadbush, for example, tend to be more tolerant of longer and more frequent meetings than other groups. Even within ideologically homogeneous groups, however, participants may have different ideas about how meetings should be conducted and how often they should be held.

The following discussion looks at four types of meetings: (1) general assemblies, which are relatively large and open to newcomers; (2) "open space" meetings, which are largely unstructured and open to input from participants; (3) collective meetings, which employ consensus decision-making procedures; and (4) governing committee meetings, including less formal Coordinating Committee meetings and more formal Executive Committee meetings. Table 7.1 provides a summary comparison of the features of these meetings and their consequences for the groups. These categories do not exhaust all of the different types of social movement meetings that I observed, which also included working groups such as MP's Outreach and Strategic Messaging committees, Sierra Club action team meetings, and Protect Our Parks (POP) coalition meetings. Special forums and strategic planning meetings are discussed in Chapter 8. In this chapter, my goal in describing some of the meetings I observed is to show how the structures and cultures of groups help or hinder their ability to meet organizational challenges, such as managing numbers, making decisions, attracting participants, and building relationships. These

TABLE 7.1 *Social movement meetings*

Types of Meetings	Features	Effects
General assemblies (Marcellus Protest)	• Relatively large and open meetings • Agreements on meeting process • Rotating facilitators • Ad hoc guidance from participants • Working group discussions and report backs	• Diverse backgrounds and ideologies of participants • Uneven organization and preparation • Potential for both enthusiasm and frustration • Follow-up requires working group action or ad hoc leadership
Open space meetings (Transition)	• Relatively large and open meetings • Usually no predetermined agenda • Small group discussions • Volunteer facilitation	• Open atmosphere and culture of participation • Diverse participants • Lack of continuity in participation • Creative ideas; new ones at each meeting • Unclear purpose of some meetings • Lack of follow-up • Uneven facilitation
Collective meetings (Shadbush)	• Small meetings of six to eight core members • Membership by invitation, screening for ideological agreement • Attention to process, use of rituals and procedures • Collective decision-making • Rotating facilitation • Extensive discussion of strategy and debriefs on actions	• Homogeneity and difficulty expanding • Close relationships among members • Creativity and commitment • Sometimes lengthy meetings • Uneven facilitation and organization • Extensive commitment required of core members
Executive Committee (Formal governing committee of Sierra Club Allegheny Group)	• Elected membership with two-year terms • Attendance by ten to fifteen people, including past officers and committee chairs as well as current ExCom members	• Stable leadership with clear process for replacement • Friendly relationships among core members • Organizational memory available • Regular organization and agenda

(*continued*)

TABLE 7.1 (*continued*)

Types of Meetings	Features	Effects
	• Field office reports from national or state staff members • Frequent presentations from representatives of other organizations • Use of parliamentary procedure • Quorum of five out of eight ExCom members • Facilitation and note-taking by officers	• Standard repertoire of discussion and action • Opportunities for connecting with other groups and alliances in the community
Coordinating Committee (Informal governing committee of Marcellus Protest)	• Shifting membership of volunteers; six-month commitments requested • Quorum of five participants for votes • Rotating facilitation, often informal discussion	• Difficulty in getting representatives of committees and groups to attend, and difficulty in replacing members • Large amount of work done by a small number of volunteers

organizational dynamics are in turn related to strategic capacity and the ability to engage in collective action.

Most of the relatively large, open meetings that I observed were held by Marcellus Protest. These meetings typically attracted more than thirty people, which may not seem very "large," but many social movement meetings attract fewer than ten people. Larger meetings often face the challenges of coordinating diverse groups of people who don't know one another well and who may have different political views and experiences. Marcellus Protest general assemblies held before November 3, 2010 were focused on organizing the demonstration, whereas those after that date had to build an organization and ongoing movement. Transition Pittsburgh experimented with open space meetings in trying to build a very different type of movement focused on creating sustainable communities, based on an international model. Shadbush operated as a small, radical collective. The Sierra Club ExCom operates as the governing body of the Allegheny Group, while the Coordinating Committee was the default governing committee of MP.

MARCELLUS PROTEST GENERAL ASSEMBLIES

The first public meeting to organize a Marcellus Shale Protest was held on September 3, 2010 at the East Liberty Presbyterian Church, which is a popular place for larger social movement meetings because the liberal congregation allows nonprofit groups to book rooms in its large, Gothic-style church at no cost. I found out about the meeting via an electronic announcement sent by the Thomas Merton Center, the local peace and justice center. Having previously seen *Gasland*, I had some understanding of the issue and was very interested to hear about what was being done locally to combat fracking. Around forty people attended the meeting – I counted thirty-four when we introduced ourselves, but more people sprinkled in late, making it hard to get an exact count.[1] Attendees were diverse in terms of movement experience, social class, and age, ranging from college age to elderly. Several people mentioned seeing *Gasland* when they introduced themselves. The organizers of the meeting had prepared an agenda, which they passed out, and Patrick Young, an experienced activist and facilitator, led us through the meeting. He introduced the first agenda item as "agreements about how to interact together in meetings," saying that this "may seem a little hokey, but it's important." These points were also written on a large flip chart and he went through each: step up, step back (talk more if you tend not to talk, talk less if you tend to dominate); stay on topic, stay brief; make "I" statements (speak for yourself only); be polite, be respectful; don't interrupt each other; give the meeting process a chance. When he asked for our input, someone asked if we would keep stack, and a volunteer agreed to do so though it proved unnecessary at this meeting because it was rare for multiple persons to raise their hands to speak. Patrick also noted that Mel Packer, one of the organizers of the meeting, had volunteered to be a "vibes watcher," but this also proved unnecessary since it was a very upbeat meeting.

The path of an emerging group can be set early on, seemingly by chance, making it difficult to change group culture and structure (Blee 2012). But my observations show how experienced organizers and flexibility on the part of participants can prevent a meeting from getting bogged down as a group is getting started. One example of this occurred when a young woman asked if we should talk about how to make decisions in the group. This created momentary uncertainty until she added, "or am I getting ahead of things?" Patrick then said he didn't think we'd be making decisions at this meeting, and Mel said, yes, she was getting ahead of things, so she agreed it was fine not to talk about this tonight. This was fortunate because it probably would have dragged down the meeting at the beginning to have to agree on decision-making procedures.

As this meeting proceeded, direction from the facilitator was important in keeping the meeting on track: After getting past the issue of decision-making, Patrick asked first for group introductions from representatives of organized groups such as the Lincoln Place Action Group, What the Frack, Shadbush, and

[1] The following description of the September 3, 2010 meeting, including quotes, comes from my fieldnotes.

others. After we exhausted the representatives present, others started talking about groups in various places that they knew about, but Patrick cut this off after a few minutes, saying there are hundreds of such groups in Pennsylvania, New York, and West Virginia, but we do not have time to mention them all. We then moved on to individual introductions, followed by a description from Mel of the action to be planned. Then Patrick asked us to suggest working groups needed, which he recorded on the flip chart. The resulting list was lengthy and complicated, and for a few minutes, it looked like there would be a lot of confusion as to what working groups would form. However, Patrick stepped in and suggested that everything could fall under four committees: outreach, logistics, media/visuals, and fundraising. We then divided into groups based on our interest in these committees, and each group later reported back to the larger group. Patrick walked around to each group, asking questions and offering a few comments that seemed helpful but not intrusive. Before the meeting adjourned, we debated when to meet next, with some arguing for weekly meetings and others suggesting we meet in two weeks, before we agreed on Patrick's proposal that the working groups meet next week and the larger group in two weeks.

As we walked out of the two-hour meeting together, Mel commented to me, "It was a good meeting, wasn't it?" and I enthusiastically agreed. But what exactly made it a "good meeting"? Mel noted that it was great having so many young people there, that "we need them." I added that Patrick was a very good facilitator and he agreed. The facilitation was certainly important; Patrick moved the meeting along when necessary, and he hit the right balance between soliciting input from everyone and providing needed direction at potential stumbling points. He was also skillful in introducing the cultural norms of the meeting in a casual, nonpreachy way, which helped a diverse group feel welcome. While skill was needed to manage the group, diversity was also important to the success of this meeting – and to the demonstration that eventually followed. As Mel observed, the large presence of young people was an invigorating contrast to the many all too familiar meetings of environmentalists and other progressives consisting only of older people. In addition, the meeting attracted more than the "usual suspects" who show up at many progressive movement meetings in Pittsburgh; the number of people attending and the new faces among them were very encouraging. While there was little racial diversity, there was class diversity and an array of dynamic styles. For example, Mark Schneider from the Lincoln Place Action Group introduced himself as a ten-year army veteran and fire fighter who also worked a second job. He told us that he was "not polite, not politically correct," and he declared passionately, "I will fight this, I will stop it in my neighborhood." A wide variety of resources were on offer from attendees, such as an airplane pilot who took pictures flying over fracking sites, artists, people with media, website and computer skills, and environmentalists with deep knowledge of issues and connections to various organizations. The group felt like one that was going to really do something, and everyone seemed excited to be part of it.

Meetings that convey the potential for effective action help to launch campaigns, and Marcellus Protest got off to a good start with this meeting. Nevertheless, the emergent group faced ongoing challenges at seven more general assemblies and numerous committee meetings held before the November 3, 2010 demonstration. The meetings were attended by some newcomers each time and facilitated by an array of volunteers with different levels of experience, most of whom created agendas in advance with help from others. At the second meeting, on September 17, 2010, we tackled the issue of how to make decisions at the general assemblies, which took some time and might have created conflict early on. Fortunately, a few experienced activists, including the meeting facilitator, sent out some suggestions about group procedures and decision-making by email in advance of the meeting, saving time at the meeting. They suggested that working groups be allowed to operate autonomously, but should be open to suggestions, and that the large group should try to come to consensus but make decisions by simple majority vote if consensus could not be reached and two-thirds of the participants called for a vote. There seemed to be agreement about this, and as far as I know the voting procedure was never needed. As Mel commented in supporting the rule, "Based on the time we have spent discussing decision making, I think it is clear that this is not a group that tries to ram through decisions" (fieldnotes, September 17, 2010).

The early general assemblies also took up other difficult issues. The media committee (which became separate from the art committee after the first meeting) wanted to do weekly press releases, which raised the question of what could be said in the name of the group. The larger group decided that we should come up with demands and talking points, which could guide the working groups. However, there wasn't enough time to do this at the September 17 meeting, where the item had been placed at the end of the agenda. This meant bumping it to the next week, when a number of people would be out of town for Appalachia Rising, a mountaintop removal protest in Washington, DC, and some argued for having the meeting on a different day of the week. This proved complicated as others couldn't make alternative days, so we decided to keep the meeting as scheduled, which meant that a smaller group of people was present. At the meeting on September 24, 2010, seventeen people participated in wordsmithing a mission statement, which was unsurprisingly difficult but nevertheless got accomplished. There was no time at that meeting to tackle demands and talking points, which we decided to schedule for the next meeting and also to start working on via the Google Group beforehand. This was a complicated and not entirely satisfactory process, but some progress was made on clarifying the group's positions.

There were also some sticky tactical issues dealt with in the general assemblies. When Loretta and Ken Weir proposed a demonstration on November 4, the day after the big event, at the site of a municipal waste facility to protest the disposal of wastewater generated by state fracking operations, there was concern over whether or not the protest would involve civil disobedience.

Marcellus Protest agreed to endorse the action only after assurances were made that it would not involve arrests, which the group felt unprepared to deal with. Following the Appalachia Rising protest, a participant told the group that some radical environmentalists who were coming to Pittsburgh for the protest from out of town wanted to know if there would be a place for civil disobedience. This set off a long discussion that was resolved with the decision that the November 3 demonstration would be completely legal, but that anyone who wanted to participate in civil disobedience could organize such actions for November 4, though not in the name of Marcellus Protest.

Both logistical and substantive concerns might have tripped up the emerging organization and movement. It was a culturally and politically diverse group that included anarchists, liberals, and others, and while this made the group exciting, it also created difficulties. Patrick commented in retrospect: "It was a chaotic and kind of undisciplined scene. Anybody could walk in the door at any time. There was not necessarily a lot of continuity from meeting to meeting" (interview, April 12, 2015). With a wide range of facilitators of different experience and skill levels, the meetings did vary greatly in how well organized and run they were. Most were kept to two or two and a half hours (usually due to restrictions on use of the room), but time often ran out before the agenda was completed. However, there was a great desire to make things work, and the goal of organizing the November 3 protest kept the group focused. When the facilitator of one meeting had to leave after two hours but others were eager to continue the meeting, another participant quickly volunteered to take her place (fieldnotes, October 29, 2010). There were also occasions when experienced organizers stepped in when the meeting was going well over the scheduled time to say, as Patrick did, "I think we're done here tonight" (fieldnotes, September 24, 2010). Knowing when to end a meeting – not cutting it off when people are excited to continue, but not letting it continue when participants are tired and the meeting is in danger of becoming unproductive – is an important skill.

Experienced activists also played an essential role in preparing the groundwork for meetings by drafting agendas and sending out suggestions for how to proceed regarding decision-making procedures, ideas for a mission statement, etc. Of course, the availability of an email list was critical, and David Meieran, who designed the original website, also set up Google Group email lists for the main group, the Coordinating Committee, and other working groups, some of which remain in use to date. During the meetings, individuals occasionally took it upon themselves to summarize what had been agreed upon or stated so far, which helped create consensus and move the meeting along. In addition, individuals sometimes exhibited flexibility or the willingness to back off their positions for the sake of consensus. For example, a young woman who suggested we refer to "raising awareness" rather than "protest" in the mission statement cheerfully backed down when she saw that others disagreed, saying, "I think raising awareness often involves protest ... but it's okay, I'm over it!" (fieldnotes, September 24, 2010). At the same meeting, Patrick withdrew his suggestion to include the phrase "stop drilling" in the mission statement because he said there was clearly no consensus on that phrase. In this way, the group was able to get past potential sticking points and advance its work, though such compromises and decisions might have subsequent impacts on the group. For example, as described in Chapter 4, the use of the term "protest" continued to be debated as the group sought to expand in more conservative, rural areas. And some activists decided to work in more ideologically homogeneous groups after the November 3 demonstration rather than continue to compromise politically in a more diverse group.

Just as decisions made in the group affected future paths of action, the influence of earlier campaigns or actions could be detected in MP general assemblies. I observed a few clashes between individuals that were likely carried over from previous organizing experiences such as the G20 protests in Pittsburgh in 2009. For example, participants in a failed G20 climate camp could be observed expressing impatience with one of the persons involved in the previous organizing debacle. In another instance, an individual who was seen as creating potential risks for protesters by his unrealistic suggestions during the G20 campaign was disdained by some of the radical activists in Shadbush. His involvement in the November 4 demonstration to protest illegal dumping of fracking water was probably one reason that concerns about potential arrests were raised in advance of the demonstration. Nevertheless, such frictions between individuals were not extensive, and there were also positive relationships formed in previous movements and campaigns, including the G20 protests, that carried over to Marcellus Protest (Wood et al. 2017). Successful organizing experiences also benefitted MP. For example, a number of young radicals had participated in a successful alliance called the Pittsburgh G20 Resistance Project (PGRP), which served as a model for some subsequent organizing (Staggenborg 2015). In particular, the model of relatively autonomous working groups alongside a general assembly was suggested by PGRP participants. The PGRP had developed talking points to guide their working groups and media spokespersons while allowing them to make decisions independently, a successful process that former PGRP activists also advocated for MP.

The meeting held after the November 3 protest, on November 12, 2010, was an important turning point for the emerging organization, and it experienced both continuity and change. In creating a structure, the model of largely autonomous working groups and general assemblies used before the November 3 protest was more or less imported to the ongoing MP organization. In addition, the group decided to form a Coordinating Committee, which had the potential of providing leadership to the organization. Before November 3, the group had no official leaders, but some experienced activists stepped in to guide the group in various ways, such as planning agendas, making suggestions in advance of meetings, and helping to move meetings along. Now, the group had to figure out if more leadership was needed and how leaders might be selected and organized. Some participants on November 12 proposed a "steering committee," perhaps with elected leaders, while others envisioned a more decentralized organization with a "coordinating committee" that helped connect the various subgroups and facilitate actions devised by working groups. With a number of people already leaving the meeting, and time running short, the group quickly chose the latter model, and a number of people volunteered to serve on the new committee, at least half of them newcomers to MP. At the end of the meeting, the new working groups huddled together to exchange contact information and set meeting dates (fieldnotes, November 12, 2010).

At the next general assembly on January 8, 2011, which was attended by about twenty-one people, the Coordinating Committee and working groups, including Outreach, Action, Internet, and Fundraising, reported on their activities. The CC report raised questions about its structure and underlying differences among members. Although it was initially thought that the different working groups, as well as member groups such as Shadbush, would all send representatives to the CC, not all were able to do so. Outreach was by far the largest working group, and one member raised the issue of having multiple representatives from that group while noting that some of the other groups were really "imaginary" insofar as they rarely met or had only one member. Another issue was whether or not groups had to send the same representative to each CC meeting. It was not clear to everyone exactly who was on the CC. One person devised a form for requesting money, and some members wanted the CC to follow a set procedure in allotting funds, while others preferred a more informal process. The CC was not getting minutes from all of the working group meetings, as some argued it should. Someone suggested that the CC should have a quorum before making decisions. A representative from the Butler County group said he didn't think the CC should make decisions for them, and others agreed that Marcellus Protest was an "umbrella group" and that the CC did not have "power" over its members. There was some disagreement over how structured the CC should be and how much leadership it should provide, with some preferring a real steering committee while others wanted to avoid concentrating any power in the CC. Few of these issues were resolved at the meeting, but it was agreed that the CC would plan and call the general assemblies.

After these early meetings, the general assemblies were held irregularly, with different formats and varying atmospheres. Even very long meetings could be positive and energetic if they conveyed the feeling that a lot was happening in the movement or if participants exchanged new ideas. This usually happened when enough people attended the meeting, and members of community groups from areas near Pittsburgh, such as Butler County and Beaver County, were very important as regular attendees of MP meetings and forums. Between November 17, 2012 and July 20, 2013, the MP Coordinating Committee organized five "Frack Forums" that included potluck meals, panels of organizers, presentations, films, and strategy discussions. The most successful ones shared stories of what movement activists were doing, giving everyone hope that change was possible. At one forum, Aaron gave a presentation on various movement activities such as the Tour de Frack and the "Stop the Frack Attack" protest in Washington, DC in the summer of 2012. In connection with the International Day for Action Against Fracking, he said, "We're not alone, we're part of a big movement" (fieldnotes, November 17, 2012). Meetings such as this were important in conveying this feeling and creating and maintaining ties among people in the environmental movement community.

TRANSITION OPEN SPACE MEETINGS

Transition Pittsburgh tried to use some version of the open space model of organizing advocated by the international Transition organization (Hopkins 2008, 2011) for many of its meetings. In this model, meetings are loosely run without a predetermined agenda to allow participants to talk about issues of concern. The open space meeting is intended to generate ideas that will lead to practical projects. A number of Transition Pittsburgh's meetings and events were well attended by people who found the organization's approach highly appealing. That is, they liked the idea of building sustainable neighborhoods and communities and creating or strengthening ties among residents working to reduce or eliminate reliance on fossil fuels. Some people who came to Transition meetings were already working on projects such as community gardens, organic farms, permaculture, and renewable energy, and others were eager to become part of such endeavors. Transition meetings could be enthusiastic and well-attended events that generated a lot of discussion. The group also ran a film series, showing films about environmental issues and projects followed by discussions. Like MP, Transition's open meetings and events attracted a fairly diverse collection of people with different movement backgrounds, a wide age range, social class variety, and some racial diversity as well.

Transition meetings of various sizes and types generated multiple ideas for activities and projects, and they provided educational and networking opportunities for participants. Entrepreneurial individuals willing to spearhead projects were able to make contact with others who might support them at Transition events. For example, a newcomer to Pittsburgh interested in aquaponics reported on his experience at a Transition event:

> I wanted to connect with people. And I knew there would be a community there of people to at least connect with and to get involved with sustainability issues. And I didn't really know what would happen to it. And at the very first meeting in typical Transition fashion, they say, "okay, now's the time to say what you want to work on!" So everyone stands up and says, "okay, I want to work on this" and then at the end of everyone declaring what they want to work on, you break out into groups to see how many people are interested in your ideas. So I said, "hey! I want to work on aquaponics in Pittsburgh." And like four or five other people gravitated around me and said, "Yeah, what

is this aquaponics? I have heard about this. I want to do this, too." And that's how it kind of helps. So, Transition was helpful in me getting to meet people here in Pittsburgh. (Interview, May 23, 2013)

Others who were already involved in sustainability projects were also able to find additional supporters at Transition meetings, and several projects such as Schwartz's Market and the aquaponics project became loosely affiliated "Transition projects." However, in the absence of entrepreneurs such as the founders of these projects, Transition lacked the capacity to follow up on most of the ideas generated at its meetings.

One of the difficulties faced by Transition was that a different set of people often came to meetings from week to week. Similar to the complaint quoted above about early MP assemblies, anyone could show up, and there was little continuity from meeting to meeting. As one of Transition Pittsburgh's founders described:

> When we first had Transition meetings, it was every Tuesday ... and open to anybody to come talk about Pittsburgh and Transition Pittsburgh and making things happen. And it turned into Transition 101. We were having the same conversation every week. We'd host the meetings and people would come and we'd take notes because everyone had lots of ideas. And then we'd do some leg work to get that idea moving, and then they don't show up the next meeting. A whole bunch of new people show up. And they want to know what Transition is, and what they can do to be a part of Transition Pittsburgh. It was like half were coming to try to get things done, and the other half – it was more like an open-door thing. But you can't require people to come. Then it's not participatory. (Interview, November 23, 2012)

While early leaders may have tried to "do some leg work" to follow up on ideas offered at meetings, more commonly there was little or no action on the ideas offered at open space–style events. At events such as film screenings, when attendees asked for ways to get involved, they were typically directed to various websites or told to seek out activities such as community gardens.

Transition Pittsburgh provides a good example of the problems that grassroots groups face when they fail to create an effective leadership team. While a number of entrepreneurs ran their own projects in the Pittsburgh community, Transition was unable to direct projects and develop the organization in the absence of leadership. Some potential leaders ended up lacking the time to contribute to the organization in order to translate ideas into projects. Some were inexperienced organizers who tried to initiate projects but had little success in doing so. Others did not think leaders should organize activities, which they thought would emerge more organically, and some did not feel that continuity from meeting to meeting mattered. Thus, interesting ideas proposed at Transition meetings were rarely followed through on, and at subsequent meetings, entirely new ideas were often proposed.

The structure of meetings contributed to some of the difficulty in following through on ideas and initiating actions. At one meeting, a drop-in format allowed people to arrive for a potluck meal over a period of four hours, meaning that not everyone was present at the same time to have a coherent meeting (fieldnotes, April 16, 2011). At another meeting where strategic planning was to be discussed, the young organizer allowed two hours for informal discussion before beginning the formal meeting. Although only about fifteen people were present, he proposed that we divide into three or four smaller groups. When an older participant objected to dividing into small groups since there were not many of us to begin with, the organizer persisted, saying that the point of "open space" was to have discussions with five to seven people. By the time the small groups were called back to the larger group, an hour and a half into the "formal" part of the meeting and three and a half hours after its start, the meeting was fizzling out (fieldnotes, January 4, 2012). In general, the style of Transition meetings was very casual, frequently including social components and activities such as dancing, and their purpose was not always clear. However, warm feelings were shared at the meetings, and members frequently hugged one another in greeting. Often, meetings ended with a circle of people holding hands and then raising their joined hands after a few minutes of silent reflection. Although the culture of the group might not appeal to everyone, it was always inclusive and welcoming to newcomers.

Organizers and participants did come to recognize that Transition lacked a structure for implementing the ideas generated at open space meetings. One person said in an interview that "we managed to have some meaningful, real conversations with people about things that [we] were caring about in the community" but that the group was unable to convert the conversations to actions (interview, July 21, 2013). An organizer regretted that many of the meetings "didn't have any clear goal or any clear actions come out of them" (interview, April 2, 2013). A young black woman who attended multiple meetings, and might have connected Transition to other young African Americans, said, "The only problem is that when you do go, when you do show up to one of these events, or one of these [meetings], it ends up being, like, there's absolutely no action you can take" (interview, July 18, 2013). In recognition of this dilemma, Transition held a long meeting in July 2012 to discuss the structure and potential strategies of the group, but, as Chapter 8 describes in detail, it did not solve the problem.

While Transition Pittsburgh was part of a national and international network, and groups in the United Kingdom and other parts of the United States have successfully created initiatives, there was not enough diffusion of organizational learning in Transition. The national Transition trainer who conducted several Trainings for Transition (T4Ts) in Pittsburgh suggested that the young activists in Pittsburgh may have misunderstood the open space concept. Perhaps, she said, "I didn't emphasize enough that open space is about getting practical work done. It's all about getting concrete results" (interview, May 9, 2014). At the T4T I attended, there was strong coverage of the issues of climate change and peak oil, and there was much discussion of our visions of sustainable communities, individual feelings and internal transformations, and Transition principles such as its positive vision and inclusiveness, but little focus on practical things to do and ways to build the movement in communities (fieldnotes, June 5, 2010). *The Transition Handbook* (Hopkins 2008:168–9) and *The Transition Companion* (Hopkins 2011:220–1) each include a description of how to conduct open space meetings, but each book repeats the four guidelines that "whoever comes are the right people"; "whatever happens is the only thing that could have happened"; "when it starts is the right time"; and "when it's over, it's over." These suggestions do not encourage the kind of planning and leadership that might make for a more productive meeting. The model does encourage participation, which is attractive to many Transition members, but in Pittsburgh it also created frustration at the lack of follow-up and results.

SHADBUSH COLLECTIVE MEETINGS

Meetings of the Shadbush Environmental Justice Collective were small, typically attended by six to eight core members of the collective. The membership changed over time, though several members were involved for most of the group's 2010 to 2014 life span. As the gaps in Figure 6.2 illustrate, Shadbush also went through a few "boom and bust" cycles – meeting regularly for months, and then failing to meet for a while. The upswings in activity occurred either in response to activities in the larger movement or because new members joined the collective. After organizing in June 2010, Shadbush was quite active for about nine months and then picked up again after a gap of several months. The first meeting I attended was in August 2011, and the group had not met for a while when I first asked about joining a month or so earlier. However, two other recruits, a young couple who had recently moved to Pittsburgh after being active in radical environmental groups elsewhere, also expressed interest in joining, and the three of us attended our first meeting together. At that meeting, Jessica and Ben recounted the history of the group for us and explained that the collective was at a "refocusing point" after not meeting for a few months. Ben noted that they had recently spent time on "group process," which took a lot of energy, and another member urged, "bear with us!"

Ben asked us if we had seen the points of unity, which we all had, and he explained that Shadbush has been "a closed group" with membership by invitation only and that there is a "process of getting to know new members." After receiving feedback from some former members that the meetings were "hard to follow," the group wanted to be sure to have good facilitation of meetings. Ben told us that Shadbush operates by consensus with a facilitator who rotates with each meeting. We talked about the practice of "keeping stack," which may not be needed for small meetings, but a few people thought should be used if six or more people were present and could always be requested by anyone at a meeting. Ben said they did not usually use a "strict set of hand signals." Nevertheless, another member went over the main signals, such as wiggling one's fingers in agreement and putting up one finger to provide a point of information, for the benefit of anyone not familiar with them. She also noted that they used to always set the next meeting time during a meeting and proposed this as a good practice to reinstate after a period without meetings. Ben suggested that we should also work on the agenda for the next meeting, which we did before adjourning. As newcomers, we were asked for our feedback on this introduction to Shadbush, which we agreed was useful, and the group then came up with the idea of doing an "information session" for prospective members outside of the regular meeting time to help recruit new members without repeating information during a meeting. This was later done once and reportedly went well, attracting several potential recruits, but was not instituted as a regular practice due to lack of capacity, and no one joined Shadbush as a result of the first session.

During "boom" periods, Shadbush meetings were frequent, scheduled every two weeks, and they almost always included a facilitator and a note taker. Usually, someone took the initiative to draft and post an agenda online before the start of the meeting. However, the agenda was always discussed and expanded at the start of the meeting, and this was sometimes a lengthy process. The agenda frequently included sometimes time-consuming reports not only on Shadbush activities but also on various events that members had attended, such as Earth First or Mountain Justice gatherings. There was a tendency to put a lot on each agenda; as one member reflected about meetings, "Sometimes they were long and sometimes we probably had stuff on the agenda that we really didn't need to talk about as a group" (interview, April 9, 2015). Meetings often began fifteen minutes late, as we waited for expected participants to arrive, and they sometimes went over two hours, which was usually the amount of time allotted. Problems such as tardiness were recognized and discussed; at one point, it was decided to have a social hour prior to the meeting so that members came on time, and at another time the group jokingly assigned "punc points" to those who were punctual. Shadbush also added features to its meetings over time, such as a routine "check in" at the start at which we went around the circle and reported on how we were doing and what was going on in our lives. Members seemed genuinely concerned about one another, and this practice made me feel like a member of the group, as I shared things about my life along with the others. At a few meetings, there was also a "check out" to determine how members were feeling about the meeting process, which Shadbush took very seriously.

As other scholars have described, movement organizations have different emotional dynamics, which affect participant commitment and group actions (Gould 2009; Summers-Effler 2010). Participation in the close-knit Shadbush collective was both demanding and rewarding. Like Transition meetings, Shadbush meetings had a distinct emotional culture, which included the sharing of personal feelings and the expression of care for one another. One Shadbush meeting, which lasted more than three hours, was held to discuss group process exclusively and, at the end of that meeting, the facilitator asked us to do a couple of exercises, including one in which we were asked to say what we valued in the person sitting to our right. Although I was dying to leave the long meeting at that point, I nevertheless found the exercise moving. Shadbush's small size and exclusive membership policy limited its capacity, but it did allow the group to develop strong relationships and a positive internal atmosphere.

In her study of German groups that used consensus decision-making, Leach (2016:55) finds that one of the ways in which they varied was in their "discussion atmosphere," which affected their efficiency in making decisions by consensus. She describes how, in some groups, participants felt pressured by features such as strict facilitation and "unspoken norms." Shadbush meetings, in contrast, were friendly affairs with informal facilitation; although some facilitators were better than others at moving meetings along, they did allow for extensive participation. The close relationships formed among Shadbush members aided the group process by allowing meetings to be relatively informal and making real conversations possible. In addition, members of the collective often socialized together after the meetings and continued conversations. As Patrick noted:

> The fact that we knew each other relatively well, and that we worked together relatively well, and could have an informal feeling meeting that was still maintaining a process, was a really good thing.... So what kept me in the group for the duration of the group was that we would have like a two-hour meeting and then we would hang out and like just talk about things afterwards.... There was a relatively seamless transition between hanging out with our friends and being in a meeting.... I see real value in that. (Interview, April 12, 2015)

In reflecting on Shadbush meetings, Robin compared them to those of other groups she was familiar with:

> Shadbush seemed so much easier than any other meetings I've been to.... We didn't really facilitate them all that like clearly or strongly, you know, and it seemed very conversational which sometimes led to some people's voices getting more time, but maybe because we all knew each other so well and trusted each other, it didn't feel like there was a lot of like people trying to figure out what this was about or where their space was in the group and trying to assert their own, like, you know, this is what we should be doing ... 'cause I've witnessed in meetings sometimes where it seems like people are actually disagreeing about something in order to more assert their will and become a leader. And we didn't have that [in Shadbush]. (Interview, April 9, 2015)

Shadbush members saw themselves as "organizers" rather than "leaders," insofar as leadership implied hierarchy while organizing meant that those with more experience and skills made a point of trying to teach them to other members of the group. However, they were similar to the leaders in high-engagement groups studied by Han (2014), who invested time and energy in developing members' capacity for collective action (see Chapter 2).

Because members knew one another well, they generally felt free to debate issues and to disagree with one another without creating conflict. This is not to say that there were no power differences within the group. Some members probably felt less free than others to talk, and some who left the group may not have felt integrated into the collective or in synch with its ideological direction or internal process. Moreover, even people who were seen as having influence in the group sometimes felt their objections were not given adequate weight. For example, Patrick had strong reservations about doing a second issue of the *Shalefields Grassroots Reporter* due to the time commitment it involved, but others pushed ahead with the project despite his concerns. At the same time, quieter members noted in interviews that they felt valued in the group. Casey recalled Shadbush as "a very respectful group of people" who "allowed each other to speak and really listened" (interview, April 3, 2015). On the other hand, the group never got around to debating some issues that were proposed for discussion, such as animal rights. Casey was interested in talking about this issue, but when I asked him about it in our interview, he replied, "I can also be forceful at times and I don't want to like force my own agenda if it's not something that people in the group feel is important" (interview, April 3, 2015).

There were always more topics for discussion than there was time at Shadbush meetings, and members generally did not push items without indications that others also wanted to pursue them. At most meetings, there were at least two fairly assertive or talkative members who sometimes debated one another, but some members did influence the meeting agenda and strategy plans more than others. At one meeting, the group considered several strategic action plans, engaging in some fruitful discussion of three possibilities. As the group was trying to come to a decision, however, an influential member asked, "Why not do all of these actions?" This was a surprising choice for a group of less than ten activists, and another participant asked dubiously, "Do we want to do all of these?" However, the problems with trying to pursue all three ideas were not discussed in any detail, and several members of the group seemed enthusiastic about the plan. In this case, consensus was assumed, and no formal decision was made (fieldnotes, July 8, 2012).

Shadbush often, but not always, checked for consensus before making a decision. The use of formal procedures for reaching consensus did not seem to depend on the importance of the issue at stake but on how the meeting was conducted and who was present. The process of decision-making was often informal when the group consisted largely of those who knew one another well, and more formal when others were present. For example, at a meeting that included three newcomers, the group did a more formal "test for consensus" on the question of where the action camp should be held (fieldnotes, September 23, 2012). Meetings without facilitators seemed more informal, and some facilitators of meetings were much more consistent than others in asking for shows of consensus. For example, when Casey facilitated a meeting that considered follow-up to the action camp and the format for a retreat, he called for a "temperature check" to test agreement by using hand signals at one point and asked if there was consensus or any opposition to a proposal at two other points in the meeting (fieldnotes, January 3, 2013). Interestingly, testing for consensus by a show of agreement was part of the Shadbush repertoire, but when a relatively new member asked if we could "vote" to move on, another member politely and apologetically explained that the group doesn't typically vote and that he personally had an "ideological objection" to voting. The person who used the word "vote" looked puzzled and said she thought we voted a lot with the hand signals and wasn't that the same thing, but some of the others thought this show of sentiment was different than voting (fieldnotes, March 9, 2013). This type of cultural practice might make some potential members feel out of place in a group like Shadbush.

There wasn't always time at meetings for a complete discussion of strategic plans and the advantages and disadvantages of possible actions. At a meeting on January 26, 2012, for example, Ben reported to the group on MP's plans for a "counter-summit" on pipelines to be held on March 20, 2012 at the same time that an industry conference on pipelines and shale gas infrastructure was occurring. Marcellus Protest had ideas about actions that could be taken in local communities leading up to the counter-summit as a way for geographically distant groups to be involved in planning. Marcellus Protest would be forming working groups at its next general assembly and wanted Shadbush to participate in the planning. There was quite a bit of discussion at the Shadbush meeting about what the group might do, and Ben said he liked the fact that the event was focused on movement building and strategic planning. Another member picked up on this and said maybe we can do strategic planning at the counter-summit when there are a lot of people there. She said that the idea of focusing solely on pipelines "is not very appealing," but she was excited about talking about strategic planning and volunteered to lead an exercise at the counter-summit. The idea was considered only briefly, but a few others thought it might be a good one. After this discussion at their meeting, Shadbush members raised the idea of a strategic planning session during the counter-summit at the MP general assembly, as described in Chapter 8.

Shadbush tried to learn from its experiences and held "debriefs" or evaluations of major actions, which were frank discussions of what went right and wrong with actions and events. Shadbush also held three overnight retreats (described in Chapter 8), at which members talked extensively about their goals and created plans that were then referred to frequently in subsequent meetings. Owing to all of this discussion, members of the collective were often on the same page about a plan or topic raised at a meeting, and if their resources were over committed, they usually scaled back their plans as they proceeded. But sometimes members simply went along with others for various reasons such as uncertain feelings about a plan, lack of extensive knowledge or experience about a topic, or the desire to avoid conflict.

Shadbush's failure to have an extended discussion about Marcellus Protest and how Shadbush might work with it was consequential for the anti-fracking network insofar as it limited the collective's support of MP. Two members of Shadbush, Ben and Jessica, participated extensively in MP, especially in its first year, but members who joined Shadbush later tended to be less involved. Marcellus Protest was sometimes mentioned in pessimistic terms at Shadbush meetings, but there was no elaboration of the concerns. At one meeting, after several people agreed that relations with Marcellus Protest should be discussed, another member who was generally negative toward MP argued, "that's probably the least pressing item, to talk about another organization" and suggested that it go at the end of the agenda, when not surprisingly there was no time for an adequate discussion (fieldnotes, July 8, 2012). When asked if Shadbush might have engaged more with Marcellus Protest, one member who joined the collective in its later years said, "I feel like I became a part of Shadbush after a lot of people in Shadbush had already given up on Marcellus Protest, so most of what I heard is just how like we were more functional than them or what was wrong with Marcellus Protest and so … I don't really know." Another Shadbush member who was not around for the early days of MP said in an interview, "I was very confused about a lot of the relationship with Marcellus Protest the whole time, so I never went to Marcellus Protest planning meetings and I never was really involved."

The amount of time devoted to discussion of an issue in Shadbush was not always proportionate to its importance – though there was some disagreement about what was most important – and Shadbush sometimes "talked things to death" as one member put it in an interview. Time was occasionally spent recapping previous discussions for the benefit of members who were absent, though brief notes were also posted after meetings. Sometimes, extensive discussion over seemingly minor issues occurred, and this was occasionally recognized with humor. For example, when the group decided to host a potluck dinner for persons who might want to become supporters of Shadbush without committing to being members, there was a discussion of what to call the gathering. Patrick said it is a network for people who don't want to come to meetings all the time, so we should not call it a "founding meeting." Jessica laughed, "where we spend an hour and a half coming up with a name!" This seemed to be a reference to past organizing efforts where much time was famously spent on the group name. Ironically, however, we did spend a fair amount of time coming up with a name for the potluck event at this meeting, which ended up being close to three hours long (fieldnotes, July 22, 2012).

Shadbush had a high tolerance for frequent and sometimes long meetings because the group valued discussion of both group process and substantive issues. Although some members complained about long meetings, they were also a matter of pride; as Patrick joked when talking about the need to schedule a follow-up meeting to a three hour strategy session, "We're Shadbush, we meet longer than three hours!" (fieldnotes, May 24, 2013). The group seemed to expect to meet for at least two hours, and I noticed at one meeting that it could have ended after about an hour, but participants seemed to want to stretch it out, raising small issues that kept the meeting going for an hour and 45 minutes. Another member also noticed this and commented approvingly, "This is the first meeting in a long time where we have had time, we could have quit earlier, but we made time just to talk" (fieldnotes, May 4, 2013). At another meeting, which was unusually short because only four of us were present, a member expressed disappointment that the meeting lasted "only one hour and fifteen minutes" (fieldnotes, August 26, 2013).

Overall, Shadbush functioned well as a tight collective, but was limited by its small size and capacity. The group required a high level of time and energy from its members, including attendance at many meetings. Meetings were not necessarily efficient, but members were highly committed to meaningful discussions, and they were willing to put in the time to achieve this. The high time commitment required by Shadbush contributed to its difficulty in expanding and replacing members. When the group eventually disbanded in 2014, it was not because there was conflict in the group, but because several members were either moving out of town or no longer had time to participate due to increased work commitments. By that time, at least one key member was convinced that the group needed to become open to new members, and that this could only be accomplished by making way for a new radical environmental group, Three Rivers Rising Tide.

SIERRA CLUB ALLEGHENY GROUP EXECUTIVE COMMITTEE MEETINGS

As described in Chapter 2, researchers surveyed a sample of Sierra Club Executive Committees and found that these leadership teams were critical to effective groups but that many groups had difficulties in recruiting leaders and teaching them needed skills (Andrews et al. 2005, 2010; Baggetta et al. 2012, 2013). My participant observation of Executive Committee meetings of the Allegheny Group of the Sierra Club provides closeup data on one local group, which furthers our understanding of the strengths and limitations of ExCom leadership teams. Whereas groups such as Transition, MP, and Shadbush had ambivalent or negative attitudes toward leadership, the Sierra Club actively sought to cultivate leaders through their participation in the ExCom and other positions, but groups did not always have the training needed to create effective teams (Ganz and Wageman 2008).

The Allegheny Group ExCom is an elected body of eight that meets once a month for two hours, from 7 to 9 PM, at a regularly scheduled meeting held at the Sierra Club field office. For many years, the Allegheny Group held larger membership meetings in addition to ExCom meetings, but stopped doing so sometime in the 2000s after attendance dropped off, holding only ExCom and small action group or committee meetings instead. ExCom meetings are open to all Sierra Club members and during the main period of my observations, from

2012 to 2017, were typically attended by ten to fifteen people, including ExCom members, who are the only ones eligible to vote at the meetings. A quorum of five of the eight ExCom members must be present for a vote to take place. Allegheny Group meetings generally last close to two hours, only occasionally going over the allotted time period or stopping early by 5–10 minutes. The chair of the AG runs the meetings unless absent for some reason, in which case the vice chair or another member of the ExCom takes over. The chair also sets the agenda and distributes copies of the agenda and other reports, such as financial and issue reports, by email in advance of the meetings. The group secretary, who is a member of the ExCom, or a substitute, always takes notes for the minutes.

Barbara Grover, the chair from January 2012 to January 2017, attempted to organize meetings so as to eliminate lengthy announcements and reports. The meetings began with introductions when there were people present who did not know one another, such as new members or guests, followed by announcements of upcoming events in the community. Barb listed many of these on the agenda, which was also distributed in hard copy at the meetings, and they were typically not read out at the meetings, though participants could add information or new announcements if they so chose. Generally, the meetings got started without much delay and proceeded through regular items such as approval of the minutes for the last meeting, treasurer's report and issue chair and action team reports (all distributed in advance), and a field office report from a staff member. An "other business" item at the end of every agenda allowed attendees to add any matters of concern. Often, guest speakers were invited, or they asked to speak to the group, either for the purpose of educating the group or asking for Sierra Club endorsements or participation in various events or campaigns. For example, representatives of a wide variety of organizations, such as Tree Pittsburgh, which works to support the "urban forest" in Pittsburgh (fieldnotes, September 10, 2012); New Voices Pittsburgh, a women of color reproductive justice group (fieldnotes, March 9, 2015); and Fight For Fifteen, a group working to raise the minimum wage (fieldnotes, April 11, 2016), were invited to explain their work and how the AG might support them. Sierra Club members who did not ordinarily attend ExCom meetings sometimes asked to speak, as did a member who came to argue for more action on climate change (fieldnotes, January 14, 2013). Workers trying to bring in a union at UPMC came to explain problems at the workplace and to ask for AG support (fieldnotes, October 9, 2012). The Sierra Club is generally very supportive of unions and other progressive organizations in the city, and members often attended hearings and rallies even when the issues were not strictly environmental. In turn, they hoped to gain support for environmental issues from other progressive organizations. By inviting speakers to meetings, the AG was able to make connections with a variety of groups. In addition, the AG regularly endorsed political candidates and often invited public officials and candidates to speak to the Club and answer questions. Candidates who had already been endorsed by the Sierra Club were sometimes asked to come to a meeting and tell the group what they needed in the way of support and how members could become involved in their campaigns (fieldnotes, July 9, 2012 and August 13, 2012).

AG procedures were different from those of the other groups I observed and would likely have been off-putting to young activists in radical groups such as Shadbush that were committed to "horizontal" organization. The Sierra Club used parliamentary procedure for formal votes, including approval of the minutes, going through the ritual of asking for a motion, second, and vote. Of those present at the meeting, only members of the ExCom could vote and everyone else seemed to accept this. Others could express their views before a vote was taken, though they were not always explicitly invited to do so. If a quorum of five ExCom members was not present, votes were delayed, even on small matters such as approval of minutes. This procedure was less efficient than simply asking if there were any changes needed to the minutes or calling for a vote without the preliminary motions, but it was not overly intrusive.

As described in Chapter 3, another way in which Sierra Club procedures differed from those of the other grassroots groups that I observed was that the AG was part of a federated structure that included a state chapter and a national organization. National-level policies, such as a ban on illegal demonstrations, also had to be obeyed. This meant that some decisions had to be cleared by other levels of the organization, and AG members sometimes questioned whether or not they were following Sierra Club policy because they wanted to be certain to do so. At a meeting to plan participation in the St. Patrick's Day parade with Marcellus Protest, for example, Sierra Club members wondered if they could use "ban drilling" on signs because the national Sierra Club had not yet come out in favor of a ban on fracking (fieldnotes, January 25, 2011). In another example, when the Endangered Species Action Team (ESAT) reported plans to table with other organizations, a volunteer who was not an ESAT member cautioned that they should be careful that the positions of the other organizations did not conflict with those of the national Sierra Club (fieldnotes, April 9, 2012).

Any state-level decisions, such as endorsements of political candidates, had to be approved by the state chapter, which sometimes limited strategic options and coalition work. For example, when an ad hoc coalition of environmental groups formed to stage a large demonstration on Earth Day 2013 at the regional office of the Department of Environmental Protection (described in Chapter 3), the AG was invited to join in the planning of the demonstration and to become an official sponsor. Although the group wanted to do so, the state chair informed them that, because this was a state-level target, the state chapter had to approve participation in the action. However, the state chapter would not be meeting before Earth Day, so participation could not be approved; members of the AG could participate as individuals but not as representatives of the Sierra Club. (Later, the state chapter and the AG both approved a procedure of voting by email on urgent matters to address this problem.) The demonstration was a stimulating one that attracted many participants, and although several AG members including the chair participated, the group lost the opportunity to mobilize members for an exciting tactic outside of the group's usual repertoire.

The AG enjoyed some continuity of leadership, a resource found to aid in the maintenance of movement organizations (Reger and Staggenborg 2006). All members of the Sierra Club are welcome to come to ExCom meetings, and some issue chairs and action committee volunteers, as well as past chairs and long-time members such as Claudia Kirkpatrick, Bob Lang, Dick Neller, and Peter Wray, regularly attended meetings I observed. When a new chair took over in January 2017, Barb Grover initially served as vice chair but later stepped aside from that position, while remaining on the ExCom. She told me that she did this deliberately so that a younger member, who might later become chair, could get experience as vice chair. The presence of long-time members and past chairs was important in supporting new leadership and providing organizational memory to the group, and past chairs, in particular, were a source of advice to current leaders. Members who were active on committees and issues were regularly invited to run for election to the ExCom when seats opened up. Seats were often uncontested, though the group usually tried to find two people to run for each seat in the annual elections.

Most active members of the AG knew one another well, and meetings were very congenial. Until late 2017, the AG was lucky to have use of a field office rented by the national Sierra Club, in which tables were arranged in a square to facilitate conversation at the meetings. Experiences such as attendance at chapter retreats, which typically involved overnight outings at state parks, helped create solidarity among participants. For example, at one meeting, the group congratulated Dick Neller, a long-time member who had helped organize the recent retreat, and everyone laughed as Randy told a good-natured story about Dick's cell phone alarm going off at the retreat (fieldnotes, August 13, 2012). The state chapter presented annual service awards to nominated volunteers, including a number from the Allegheny Group. These and other accomplishments of members, such as published articles and letters to the editor, were regularly noted at ExCom meetings, and news of members, such as the death or illness of a long-time member, was conveyed and cards were signed. Shared experiences such as bus rides to events in Washington, DC, Harrisburg, or New York also created solidarity. At one meeting, a sweet moment occurred when a member who received financial support from the group to attend the 2014 People's Climate March in New York City said, "Thanks for the bus ride to New York City and back for the climate march; it was a huge experience for me. Thanks, Allegheny Group!" (fieldnotes, October 14, 2014).

Like other grassroots organizations, the ExCom struggled to complete business while avoiding dull meetings. Discussions of various financial and organizational matters, as well as strategies and tactics, occurred at ExCom meetings.

Some discussions of relatively minor matters were fairly long, such as a discussion of reimbursements to members for mileage and parking when they attended events on behalf of the chapter; decisions about how much to contribute and how to word an ad in the program of an annual event in support of the Thomas Merton Center (fieldnotes, September 10, 2012); a 40-minute discussion of plans for the picnic (fieldnotes, July 8, 2013); a long discussion of details about mailing ballots for the ExCom election, even though few members vote in the elections and this one was uncontested (fieldnotes, October 13, 2014); and a long discussion of how to reprint Trail Guides (fieldnotes, April 13, 2015). Some discussions were clearly important but difficult to have in a fairly large group, such as one about the Facebook page, which resulted in the formation of an ad hoc committee after a long and inconclusive discussion at an ExCom meeting (fieldnotes, April 13, 2015). Other important discussions were also tedious, such as a review carried out over several ExCom meetings of procedures for awarding grants from a fund left to the AG, which was administered by a separate committee that the ExCom was supposed to oversee. On this and other matters of governance, the group took its responsibilities very seriously and considered issues carefully, which did not always make for lively meetings. While informally organized groups might have chosen to skip over matters such as an uncontested election, the AG felt compelled to do everything properly, adhering to state and national rules as well as group guidelines. The ExCom did recognize that some discussions might be boring, particularly to younger members, such as interns who worked with the staff and sometimes attended meetings (fieldnotes, October 9, 2012). The chair, as well as some other members of the group, regularly tried to move meetings along.

Members sometimes came to the ExCom to raise strategic ideas, which also received some discussion. For example, one member presented his idea of having an educational "kid's corner" on the website and received feedback from the group (fieldnotes, August 13, 2012). A young law student came to the Sierra Club to discuss his interest in an anti-fracking campaign focused on showing how institutions such as law schools supported the industry. As was often noted at meetings, however, the group lacked sufficient volunteers to carry out many such projects. Claudia, a long-time member, asked the law student if he was a Sierra Club member, and he replied, "No, but I can join" (fieldnotes, August 8, 2016). He subsequently attended a number of other meetings, first to update the group on his project and later to participate in other activities, becoming a valuable member of the AG. Frequently, ExCom meetings were used to inform participants about campaigns such as the Clean Rivers Campaign, a coalition effort to which the Sierra Club contributed after its founding in 2012. Allegheny Group representatives to coalition efforts such as Protect Our Parks reported to the ExCom, and the group often discussed how the AG could contribute to strategic initiatives (fieldnotes, October 14, 2013). Randy Francisco, the Sierra Club staff member who attended most ExCom meetings that I observed through 2017, regularly provided information about the projects he was working on and other events and groups in the local environmental movement.

The culture of the AG was friendly, but, as the governing body of the group, the ExCom had a lot to do at its two-hour monthly meetings and generally did not indulge in many of the long conversations typical of other grassroots groups. While the practice of giving "report backs" on events was routine in Shadbush, it was more sporadic in the AG and seemed to depend on the inclinations of the chair. For example, there was no scheduled report back on the 2014 Climate March in New York City, and it was only discussed in comments volunteered by individuals such as the one quoted above (fieldnotes, October 14, 2014). However, reports on the 2017 Women's March in Washington, DC and the Summit Against Racism in Pittsburgh were on the agenda (fieldnotes, February 13, 2017). It was particularly difficult to have extensive strategy discussions at the ExCom meetings, and recognizing this, the group held several lengthy strategy sessions on weekends and tried to follow up on the plans made at these sessions (see Chapter 8). When the group held a "debrief" on one of these sessions, however, ExCom participants did not have a lot of comments about it, and it was much shorter than the extensive ones held by Shadbush (fieldnotes, February 11, 2013). While there was less of a culture of extended evaluation at ExCom meetings, there was not usually any conflict displayed at meetings, and participants spoke freely about their concerns. For example, AG members talked candidly about the group's lack of extensive activities and need for greater capacity (fieldnotes, November 15, 2016). If disagreements arose, the group could take a vote if necessary and end discussion, though informal consensus was often reached instead. ExCom members who failed to attend many meetings were asked to step down if they could not attend regularly so that more active members could join the board, a practice that helped make room for some younger members on the board.[2]

MARCELLUS PROTEST COORDINATING COMMITTEE MEETINGS

In comparison to the Sierra Club ExCom, which was part of an established national organization that provided a blueprint for group governance, the experience of Marcellus Protest's Coordinating Committee shows how difficult

[2] This practice unfortunately also eliminated the AG's only black ExCom member, who had previously been active in the Sierra Club but later stopped attending ExCom meetings. When asked if she was able to continue on the ExCom, she explained that she had become too busy to participate. The Sierra Club, like many other environmental groups, has had difficulty diversifying its membership in terms of race. The AG did try to support groups such as New Voices Pittsburgh, the women of color reproductive justice group, and to participate in activities such as an annual Racial Justice Summit held in Pittsburgh. The group also looked for ways to support diverse communities, such as a pollinator garden project in a low-income African-American neighborhood. While the group deliberately did not use this project to try to recruit participants to the AG, their work helped make the Sierra Club more visible in the community.

it is to create a structure from scratch in a diverse group. The CC was a different type of governing committee that faced mostly different problems than the Allegheny Group's ExCom – though both spent time at meetings discussing some tedious matters such as budgets. It was not clear what role the CC should play in MP or what role MP might play in the growing anti-fracking movement and what strategies and tactics it should pursue. At its early meetings, the CC dealt with a number of structural issues such as who counted as a member, how often and when to meet, how decisions would be made, and what role the CC should play as opposed to the general assembly and working groups.[3] After extensive discussion about its structure and authority at one meeting, the group seemed to agree with the summary offered by a member: the CC should make decisions about financial matters, create a network among working groups, seek funding, organize general assemblies, and discuss issues, while the GA would make major policy decisions (fieldnotes, March 30, 2011). However, the CC ended up as MP's main decision-making body, as the general assemblies rarely made any policy decisions. The CC also performed administrative tasks and helped MP remain active for several years, but never managed to form a real leadership team.

One major problem was getting enough people to attend CC meetings on an ongoing basis. Member organizations and working groups found it difficult to find volunteers willing to go to CC meetings in addition to all of the other meetings most attended. A quorum was first set at seven, but later reduced to a more realistic five, required to make decisions at meetings. To encourage attendance, the group established a regular, monthly meeting time, which worked for a while. To create some continuity of membership on the committee, CC members were asked to make a six-month commitment, which a number did, with some serving even longer. The CC discussed making decisions by consensus but voting if necessary, which was generally what happened. When there were not enough official members of the CC present at meetings, everyone in attendance was urged to vote (fieldnotes, February 5, 2011). Sometimes, issues were sent to the CC Google Group (which included past and present members of the committee) to get input from a wider range of

[3] McCarthy et al. (1991) explain how the American tax laws impose some structural uniformity on movement organizations, which affects strategic choices such as electoral work and lobbying. Many SMOs attain 501c3 status through a fiscal sponsor to allow supporters to make tax-deductible donations, but these sponsors vary in their requirements, which may shape organizational structure. Some structure was imposed on MP by the group's first fiscal sponsor, the Center for Coalfield Justice (CCJ), which had little experience in the role of fiscal sponsor and was very cautious as a result. The CCJ required MP to provide a detailed budget and the names of board members and encouraged voting by requiring a clear recording of decisions in minutes, which had to be submitted to the CCJ after each meeting (fieldnotes, September 22, 2011). In 2012, MP switched fiscal sponsors to become a project of the Thomas Merton Center, which was highly experienced in sponsoring projects and much more flexible in its requirements, allowing the CC to be somewhat more informal in its procedures.

people, and at times when it was difficult to get a quorum at meetings, votes were taken online, with five votes needed to proceed. (These were usually minor decisions, such as whether to endorse an action by another organization that MP would clearly support.)

CC members discussed a number of different matters at their meetings, which typically took two hours but sometimes lasted up to three hours. Agenda items were often suggested online before the meeting and finalized quickly at the start of a meeting, but some meetings had no agenda and were less organized. Some meetings had facilitators (and most had a note taker), while others were more informal. Meetings usually included reports on events that members had attended, including ones around the state such as demonstrations in Philadelphia or Harrisburg, and they often included discussions of how events planned by the CC turned out. The CC regularly dealt with administrative items such as approval of requests for financial aid, which could take time, depending on the size and purpose of the request. Requests for funding were supposed to be submitted to the CC on the form devised for this purpose in advance of meetings so that they could be put on the agenda. Although most decisions were made easily, some financial ones created conflict, in part because they reflected divisions about how the organization should be run.

At one meeting, a member showed up to request money for a project and became upset when told she was supposed to fill out the form to make a formal request. In this instance, as in others, some members felt business could be handled more informally and that formalized procedures were unnecessary. Some items were contentious, such as recurring decisions about paying a small stipend to the first webmaster, and decisions about grant applications. Some matters, such as work on the budget, provoked long and boring discussions. Particularly after other committees were no longer actively meeting, the CC was overworked, and one member noted that it was not only a lot of work to participate but often a "thankless" job (fieldnotes, August 11, 2012). On several occasions, members considered the possibility of hiring a staff member to ease the burden, but it was decided that this was not feasible, both for financial reasons and because it was not clear who might supervise a staff member.

In addition to discussing the structure of MP and how its working groups were functioning, the CC also planned general assemblies and considered possible strategies and campaigns that the group might pursue. It was difficult to find time to do all of this, plus the administrative work of the committee, in two-hour meetings, and a handful of people were doing much of the work. The CC sometimes tried to vary its meeting format to bring more people and ideas into the discussion. On one occasion, the CC held a joint meeting with the Outreach committee, which was attended by twenty people rather than the usual five or six who attended CC meetings (fieldnotes, June 23, 2011). (There was talk of having a joint meeting with the Action committee, partly as a way of helping to revive the committee, but this never happened.) On another occasion, the CC held a special three-hour meeting just to focus on strategic planning (fieldnotes, August 11, 2012). The committee often discussed strategic goals and potential campaigns, and it succeeded in initiating some actions, such as the counter-summit on pipelines in 2012 and the Earth Day demonstration targeting the DEP in 2013. After the Frack Forums were initiated, they were alternated with CC meetings so that CC meetings were held every other month rather than monthly, and this new arrangement made participation on the CC somewhat easier.

The CC recognized that activities were needed to bring people into the organization and expand the movement. While there was a lot of interest in the movement, and potential volunteers often contacted MP through a form on the website, the group lacked the structure to integrate them. Individuals were referred to working groups, but it was not always easy to join them. Outreach was the only working group that met regularly for about a year after MP's founding until the Strategic Messaging group was formed in July 2011. After the Outreach and Strategic Messaging groups stopped meeting, the CC was the only active MP committee, though member groups such as Shadbush, Marcellus Outreach Butler (MOB), and South Hills Area Against Dangerous Drilling (SHAADD) continued to meet. The CC agreed, as one member commented, that "meetings are not always the best way to attract new people" and that "we need to be doing things, have actions for them to participate in" (fieldnotes, March 30, 2011). When appealing actions were being planned, meetings were also more fun to attend. The dilemma was that meetings might not always be interesting, but they were needed to keep the group alive to plan actions.

While recognizing the need for more actions and campaigns to build the movement, the CC and MP suffered from a lack of capacity. To introduce newcomers to Marcellus Protest at a social event rather than a meeting, the CC organized a "meet and greet" event in 2011 at the home of a member at which MP representatives talked about group activities and how to get involved. While the CC considered this mobilizing tactic promising, it was not repeated because there were not enough people available to organize such events. As a CC member commented, "It is always the same people who do everything" and "everyone is stretched" as a result (fieldnotes, May 9, 2012). At one meeting, we discussed the fact that the Action committee was overly reliant on one leader to organize actions, and that there were not enough people to plan general assemblies (fieldnotes, May 2, 2011). After it was suggested at a number of meetings that MP needed a volunteer coordinator, a member of the CC finally volunteered for the job (fieldnotes, September 1, 2011), but she was uncertain about how to integrate newcomers. In this case, a recruitment committee, rather than an individual coordinator, might have been more effective in generating ideas. Another suggestion was that MP might create a new fundraising team consisting of organizational members (fieldnotes, September 22, 2011), but this idea, like many others, was never implemented. The main reason for lack of follow-up seemed to be lack of capacity.

Marcellus Protest failed to solve some structural problems that might have increased its resources and capacity. One reason for this was that MP consisted of a fairly diverse group of people who had different ideas about how to proceed, and the CC reflected some of these differences. Some members of the CC resisted any type of hierarchy in its structure, or that of MP, while others wanted a more traditionally structured organization with leaders who were authorized to take action and accountable to the group. Early on, some members advocated making decisions by consensus or, failing that, a super-majority vote. Some wanted clear rules about process, such as requiring new members to attend two meetings before being allowed to vote. Others preferred a more informal structure, allowing whoever showed up to vote when necessary, and simply discussing matters among participants. In fact, this was often what happened, particularly when fewer people came to CC meetings. As Mel commented to me after a small but productive meeting that included people who sometimes disagreed with one another, "democracy is easier with only five people!" (fieldnotes, April 12, 2011).

A member who preferred a more traditionally structured type of organization suggested that the general assembly could discuss issues, but then the leadership in the CC should make decisions and organize campaigns (fieldnotes, August 11, 2012). Expressing frustration with the decision-making process in MP, she said that some people originally wanted to create a more formal structure after the November 3, 2010 demonstration, but others did not want this, and consequently, there is "no coordinated effort because there is so little organization." As a result, she said, she decided that "I will go and do my own thing" outside of MP, and she suggested that MP could support people doing things and connect people to one another rather than creating its own projects (fieldnotes, November 16, 2011). At a subsequent meeting, however, the same activist expressed hope that MP could get back to doing larger-scale events after having "lost a lot of steam" after the successful November 3, 2010 demonstration due to a lack of structure (fieldnotes, December 14, 2011). Significantly, she was far from the only person to express multiple views about MP's role; many supporters were uncertain about the best way to structure the organization and continue the movement.

One of the dilemmas that the CC struggled with was the desired frequency of meetings. At a meeting where strategic ideas from the previous general assembly were discussed in the context of planning the next one, some members argued that there was too much talk and too little action. Diane asked: "Why not build on what we talked about last time – having a school group, a farm group – appoint a facilitator and follow through. We have the problem of a lot of good ideas, but no follow through ... We need to stop meeting and start doing more." Loretta argued: "People want to do something, they want direction. When people get muddled down in meetings and committees, you are preaching to the choir. We need definitive direction – for example, dates for PTA meetings [for a suggested school campaign] – we need something concrete" (fieldnotes, September 22, 2011). At other CC meetings, the group discussed the issue of meetings versus action differently. In discussing how to keep people involved, members discussed whether or not there had been too many meetings after the November 3 demonstration, with most agreeing that was not the problem. Diane said regular meetings are needed to keep people together, and Jessica agreed that we need to have some meetings, suggesting that we might want to revive the Action committee. The group agreed with Loretta's suggestion that we should make things fun when we get people together, maybe have a meal together (fieldnotes, May 9, 2012).

At another meeting, Diane argued that we "need to have a presence" to keep up the movement and to have ongoing meetings. She explained how MOB, the Butler county group, attracts participants with films and speakers (fieldnotes, June 13, 2012). In another discussion, a CC member suggested that the problem with meetings was that some of them were too long, lasting more than two hours. In planning the Frack Forums, the CC did attempt to make the meetings interesting and fun, though members had different ideas about this. As the group planned the first Frack Forum, Mel agreed with a suggestion to show a film or have some other such activity rather than just discussion, saying, "We don't want the first one to be a long boring meeting!" In response, Aaron joked, "I guess I lean toward long boring meetings," and Jessica said, "I like them too" (fieldnotes, August 11, 2012). Although these comments were made in jest, there were in fact differences in the group with regard to tolerance for lengthy discussions at meetings. However, there was a recognition that meetings should not be made overly long by packing too much into the agenda.

Regarding committees, the CC debated the question of standing committees versus action committees. In discussing how to integrate volunteers into MP, some participants suggested having committees devoted to tasks such as a media response team. However, others felt that it was better to form committees around campaigns. Diane suggested that MP should "talk about things that need doing, then form committees around them." In other words, the group should think about particular campaigns or actions and then organize committees around these rather than organizing more general committees such as a media response team. She gave the example of a committee to do a media campaign around the "Toxic Trespass" legislation (fieldnotes, November 16, 2011), which was passed by the Pittsburgh City Council in December 2011 to ban corporations from dumping toxins from fracking operations. In general, the group agreed that it was desirable to have campaigns to draw people into the movement as well as to accomplish goals. The problem was that campaigns required activists to plan them (unless they arose fairly spontaneously), but to attract activists to the group, it was necessary to have things for them to do, such as working on ongoing committees. This problem might have been solved by having a mix of standing committees and action committees, but leadership was needed to develop such a structure.

While the CC did a huge amount of work on behalf of MP, it was difficult to create a leadership team that could continue to train new leaders and develop strategic capacity. Ganz (2000) argues that a diverse leadership team, with networks to outside constituents, is important for innovation. Members of the CC were relatively diverse in terms of age, class, experience, and ideology, and they had networks to some outside constituencies, such as rural and suburban communities. They did regularly discuss strategic ideas at meetings, but they had little capacity for implementing ideas, and there was a lack of structure to assist in organizing additional activists. Although some CC members served for relatively long periods of time and remained connected to MP after leaving the committee, there was a fair amount of turnover on the CC, and it was difficult to attract new members. Marcellus Protest continued to rely on energetic individuals to plan actions and events, but in the long run, individual initiatives could not compensate for a lack of structure in keeping the organization active. After the CC stopped meeting in 2013, Marcellus Protest became more of a network than an organization, maintained by its Google Groups and by the individuals who kept up (for a time) its Facebook page, website, calendar, and newsletter.

CONCLUSION

While collective actions are the public face of a social movement, they often require planning at meetings; social movement meetings are the important behind-the-scenes sites of movement activity. Frequent or regular meetings provide a sense of "movement" – that things are happening – and they are a place where supporters can make contact with the movement. Good meetings can generate enthusiasm, teach skills, produce strategies, and strengthen relationships within and across movement organizations. Meetings that are centered on planning actions and campaigns are usually the most exciting ones, and it is easier for diverse participants to work together when they have a focus such as planning a demonstration or event. Regular meetings create organizational stability and the potential for groups to recruit and train activists and leaders. But there are also many challenges associated with social movement meetings of different varieties. These include keeping things interesting, managing diversity and conflict, and avoiding overwork of activists while increasing capacity.

Comparisons of different types of meetings, held by groups with different structures, cultures, and ideology, reveal differences in the abilities of grassroots groups to meet these challenges. Marcellus Protest general assemblies leading up to the November 3, 2010 protest brought together diverse activists who were able to manage differences by focusing on the event and the excitement of a new movement. Later, MP and its Coordinating Committee had to navigate different ideas about structure and procedures without a stable leadership team or staff. The Sierra Club ExCom, in contrast, benefitted from a federated structure and elected leadership, but was also constrained to some extent by its own repertoire of tactics and organizing. Shadbush created involvement and commitment with meetings that focused on building relationships as well as generating activities and projects, but the collective never figured out how to increase its capacity by becoming a more inclusive organization. Transition meetings generated enthusiasm but suffered from a lack of follow-up; the open space meetings might have been more productive if they were focused on particular projects.

Meetings called to plan actions and campaigns such as Protect Our Parks are probably the best at managing diversity and spurring coalition work. Meetings are important in generating ongoing movement activity, but grassroots groups often find it difficult to sustain the commitments of participants over time. The meetings I attended were interesting when participants brainstormed ideas and shared their passions with one another. They were also enjoyable when members laughed together and showed their appreciation for one another. However, all of the groups I observed had some meetings that were at least partly boring or tedious. In fact, I was continually amazed at how willing many activists were to conscientiously slug through important but sometimes wearisome tasks such as budget reviews or discussions of group process. Of course, some participants were no doubt turned off by such meetings, and many activists expressed dislike for overly frequent or long meetings. Some dull or difficult content could likely be handled better by individuals or small committees outside of meetings, and all of the organizations that I observed tried to do this to some extent. However, groups committed to democratic process sometimes feel compelled to include more people in all kinds of business and decisions. Drafting agendas and sending out information in advance of meetings via email lists is often very useful, as is skilled facilitation at a meeting, but this is not always achieved. Lack of time to cover all of the topics on a packed agenda is a common problem for meetings, and skilled organizers can plan realistic agendas and move meetings along.

Diversity is an important source of innovation, and meetings are stimulating when different types of participants provide novel perspectives and offer a variety of resources. It can also be a source of conflict or difficulty as people with different experiences and political ideologies attempt to work together. Having a focused goal, such as organizing a big demonstration, is very helpful in eliciting cooperation. Experienced organizers who step in and help meetings along – sometimes by backing away from positions of their own that do not gain consensus – are important to the success of larger meetings. Differences in cultural norms might also create misunderstandings or discomfort, but discussing and explaining norms such as keeping stack or hand signals can help newcomers or outsiders feel included.

Perhaps the biggest pitfall for social movement meetings is their potential for wearing activists out rather than building capacity for the organization and the movement. In the absence of a supportive structure of committees, such as recruitment and fundraising committees, a governing committee such as Marcellus Protest's CC gets overburdened. Shadbush went through "boom and bust" cycles because the group took on so much at times that members had to step back afterward. Grassroots groups need to find ways to recruit and train new leaders and activists so that meetings are revitalized. All of the groups that I observed generated many potentially fruitful ideas in their meetings, but they only had the capacity to follow through on a few of them. In the following chapter, I look more closely at efforts to devise strategies and some of the challenges in doing so at meetings and special forums. The grassroots environmental groups in my study engaged in a remarkable range of activities, but they also faced extensive problems in trying to generate ongoing actions and campaigns.

8

Devising Strategy

Grassroots environmental organizations have mounted many campaigns and engaged in a wide variety of strategies and tactics. Movements need strategies, which include choices about tactics, frames, alliances, and targets, to mobilize supporters and have an impact on public opinion and policies. One key question for movement scholars is how strategies and tactics affect outcomes (Andrews 2004; Banaszak 1996; Gamson 1990; Luders 2010; McCammon 2012). Another question, which is my concern in this chapter, is how strategies get created (Ganz 2000; Jasper 2006; Polletta 2002). While scholars commonly view movements as a series of collective actions or campaigns (Marwell and Oliver 1984; Tilly and Wood 2013), it may be harder for movements to devise strategies and sustain campaigns than is generally recognized (cf. McAdam and Boudet 2012). In addition to assessing the effectiveness of particular strategies, it is important to take a step back to see how strategies get devised; while new actions and campaigns allow movements to grow and succeed, strategic innovation is not always easy.

Strategies are created in various settings, including the types of meetings discussed in Chapter 7 as well as in forums organized for the purpose of strategic planning. These forums vary in the extent to which they lend themselves to productive discussion of strategic options by a team of leaders or organizers. When organizations help volunteers to develop their skills, give them some strategic autonomy, and involve them in campaign-based teams, they are likely to enhance their ability to plan actions and campaigns (Han 2014:102–5). Many activists enjoy working in groups that value participatory democracy, which tend to be good at brainstorming strategic ideas (Polletta 2002). Yet we know little about how strategies are created and how forums, organizational structures, styles of interaction, and procedures affect the ability to fashion effective strategies. By looking closely at the process of devising strategies in meetings and other forums, we gain understanding of why it can be difficult to develop effective strategies and what grassroots groups might do to increase their strategic capacity.

This chapter examines how grassroots environmentalists make strategic choices in meetings and special strategic planning forums. It also looks at workshops sponsored by environmental groups to train activists in devising strategy. I examine constraints on the generation of strategic ideas that help to explain why only some are developed and implemented. Some of these constraints are external events and actions of potential targets that affect strategic planning as activists try to respond to threats and opportunities. Other constraints are organizational; strategic choices and capacity are affected by the leadership, structures, diversity, and resources of movement organizations. The following discussion of the actual process of creating strategy provides insight into both the promise and limitations of grassroots activism.

DISCUSSING STRATEGY

As we saw in Chapter 7, grassroots environmental groups continually discuss strategic issues at their meetings. They are faced with a number of problems in doing so, which vary somewhat depending on the composition of the group and the nature of the meetings. One problem is that meetings often fail to provide the best format for a good discussion of strategy, frequently because there is not enough time to discuss all of the issues involved. Particularly in diverse groups and in relatively large meetings, participants throw out a lot of different strategic ideas. While a diversity of ideas can be exciting and fruitful, there are usually many more suggestions than can be discussed thoroughly in a meeting, much less implemented; groups need some way of sorting through the ideas and figuring out how to follow up on them. The difficulties are compounded when activists disagree on which strategies to pursue, a problem more likely in diverse groups. Moreover, when heterogeneous participants are asked to discuss strategy, they often have very different ideas about what this means. Some talk about long-range goals while others focus on concrete details of suggested tactics, and there are different ideas about the type of planning that should be done for possible campaigns. These differences depend on the characteristics of participants and their prior experiences in movements and communities.

Strategic planning discussions in grassroots organizations are often very scattered, and organizers may hesitate to impose their ideas on larger groups of people at general assemblies or open space meetings. Often, there is uncertainty about how much direction organizers should provide, and the desire to be democratic and open to everyone's ideas can lead to a paucity of guidance on the part of organizers. Some groups lack a cohesive leadership team or structure for organizing strategic planning discussions. Diverse groups may lack a common culture or shared ideas about the discussion of strategy, while more homogeneous groups may have difficulty departing from familiar patterns.

Once strategies are proposed, groups often lack the structures and resources needed to generate actions and campaigns based on the ideas. When leaders lack authority to make strategic decisions on their own, follow-up is difficult. Frequently, the implementation of strategic ideas depends on enterprising individuals who take the initiative to drive the strategic plans. It also depends on the larger political context; although interactions within movement organizations are critical, external events and threats also prompt movement strategies. In Pennsylvania, a negative political climate created both threats and targets for the movement when Republican Tom Corbett, who strongly favored fracking, was elected governor of Pennsylvania in November 2010. His administration provided targets for the movement when he adopted policies and appointed officials who were hostile to environmentalism.

Leadership and Democracy

A consistent problem is that there never seems to be enough time at meetings to thoroughly discuss the many strategic ideas of participants. But why do organizers try to cram so much into meetings and why does time run out so frequently? One reason is a good one: grassroots movements tend to be committed to democracy and they want to create forums that allow for participation from everyone. Diversity is also valued, and although there was little racial diversity in the grassroots environmental groups that I observed, there was diversity by age, class, education, rural/urban residence, movement experience, and ideology. The more diverse the group, the more different ideas are offered and the more time it takes to discuss strategy. Of course, relatively homogeneous groups such as Shadbush might also engage in extended strategic planning because they are ideologically committed to doing so. Related to the desire for participatory democracy, movement organizers are often reluctant to provide too much direction to participants, and strategic planning often lacks clear leadership.

Consider some of the difficulties encountered by Marcellus Protest (MP) in strategic planning after its successful demonstration on November 3, 2010. While the initial demonstration was conceived by organizers before being presented to attendees at a large open meeting, subsequent strategies and tactics had to be negotiated among a diverse group of people with no formal leadership. At the November 12, 2010 general assembly (GA) following the protest, it was very difficult for the group to devise strategy. Not only was it a large group for a strategy meeting, but about half of the sixty or so people present were attending their first MP meeting, many after participating in the protest. Organizers distributed an agenda that included time to generate strategic ideas, evaluate the organizing of the protest, and create a structure for Marcellus Protest. With so many newcomers and so much enthusiasm generated from the march, the 15 minutes allocated for "brainstorming" strategies were clearly inadequate; the discussion went way over the time and was difficult to control.

Among the suggestions were a mixture of long-term strategic plans and specific tactics: boycotts, research on the revenue generated by fracking, organizing in rural communities, appeals to interest groups such as tourism and outdoor sports groups, a demonstration at incoming Republican governor Tom Corbett's inauguration, fundraising, yard signs, organizing door-to-door, creating media attention, forming coalitions, targeting fracking sites, protesting at water treatment plants, focusing on workplace safety issues, supporting the Pittsburgh ban, and expanding from the city of Pittsburgh to a ban on fracking in Allegheny County. A woman who said she was attending her first meeting asked some good questions: How we can prioritize all of these concerns? How would we allocate resources? How do we accomplish our goals? These were the types of questions that the group struggled with as it created a structure of working groups, coordinating committee, and general assemblies to devise and implement strategies. Some of the ideas, such as organizing in rural communities, were taken up by working groups, and some individuals took it upon themselves to do the work necessary for certain tactics, such as organizing a bus to Harrisburg for the Corbett inauguration. Other ideas were lost in the shuffle due to a lack of sufficient structure, resources, or enthusiasm (fieldnotes, November 12, 2010).

Strategic planning continued to be attempted at general assemblies. MP invited members of regional anti-fracking groups in the counties and townships surrounding Pittsburgh to the GA on May 22, 2011, and a portion of the meeting, attended by over thirty-five people, was spent discussing strategy. We began the discussion by spending time identifying topics for strategy sessions. Some of the suggestions were very specific, others more general; they included community ordinances to ban fracking, farming issues and possible use of a boycott tactic against companies that leased their land for drilling, language and talking points for a state moratorium, the structures of counties and what could be done to target them, nonviolence training and action, targeting schools and creating packets to deliver to schools, developing a toolkit for working with municipal officials, creating a how-to kit for film screenings and discussions, and message framing. Some of these ideas were consolidated before we divided into discussion groups around the topics, which later reported back to the larger group. Many good ideas were described but no conclusions reached as to how the strategic plans might be pursued before the three hours scheduled for the meeting were up and we had to vacate the room (fieldnotes, May 22, 2011). Some of the ideas continued to be discussed at CC and GA meetings, particularly when an individual took the lead, as one member did in organizing a working group on strategic messaging.

The MP Coordinating Committee recognized that it was difficult to have productive discussions of strategy in large meetings, and that there was often insufficient time for strategic planning, even in small meetings. There were, however, different ideas about how much direction the CC should provide, and what authority it had to direct strategy. At one CC meeting, it was suggested that the CC could plan small campaigns, but that bigger plans should be approved by the GA (fieldnotes, March 30, 2011). When the CC was planning a strategy session as part of an upcoming GA, David asked, "do we want to impose a direction, or maybe do some of that, but also have a more organic process where ideas are raised at the GA?" Some members wanted the CC to follow up on ideas generated at general assemblies. Some felt the CC should offer concrete proposals, while others felt there should be room for "creativity" among GA participants. Ted argued that the two approaches appeal to different types of people and that there was room for both. Elizabeth recalled that the effort to discuss strategy at the meeting following the November 3, 2010 demonstration had been "chaotic" and said that it is difficult to make serious decisions at such a meeting. Some felt that MP lacked strategic direction, and Jessica commented that, while it was true that the group lacked a long-term strategic plan, "you can't cram a good planning process into the GA" (fieldnotes, September 22, 2011).

The role and authority of the CC in formulating strategy continued to be debated. Recognizing the difficulty of deciding strategy at the GA, Mel suggested that we didn't have to take strategy matters to the GA unless they involved a change in policy direction (fieldnotes, October 19, 2011). At a subsequent meeting, Dana suggested that the CC go to the GA and present ideas and then ask people to work on them. She said, "the problem with Marcellus Protest is that we waste too much time in the democratic process of trying to organize with everyone. We need to bring ideas to the table already worked out, and then have democratic discussion" (fieldnotes, December 14, 2011). One of the difficulties, however, was that the CC was stretched thin and, by the end of 2011, there were very few active working groups that might have provided additional leadership in working out strategic ideas. The Outreach committee, which was always the most active working group, ceased meeting in October 2011, and the Strategic Messaging committee completed its work in January 2012 (though reunited for another meeting a year later to discuss ideas for workshops). There was some discussion of expanding the strategic messaging committee to deal with other strategic issues such as tactics and targets (fieldnotes, October 24, 2011), but this never came about – probably due to a lack of capacity among the small group of people trying to lead MP. Thus, it was difficult to establish a leadership team due to lack of an adequate structure, insufficient numbers of participants, and unclear authority.

A reluctance to impose leadership also showed up in other groups. When Shadbush organized with outsiders, such as participants in Occupy Pittsburgh for the demonstration against the DUG East (developing unconventional gas) shale gas industry conference in 2011, the group was careful to solicit their ideas rather than to provide strong direction. In planning a training session on direct action prior to this action, one member noted the need to "make people feel involved" in the decision to stage the demonstration rather than telling them about it as something already determined (fieldnotes, October 5, 2011). At an open meeting held at the Occupy camp to organize the demonstration, Shadbush allowed time for brainstorming, but eventually had to steer the meeting back to a discussion of tactics to be used to protest the DUG East convention. As this meeting showed, it is often very hard to balance open participation with a focus on planning for a specific event. In this case, it was especially difficult because the participants did not all know one another and had different levels of experience with collective action (fieldnotes, October 29, 2011). Planning for this event was somewhat disjointed, and some aspects of the action were more successful than others. Some outside participants failed to come through on their commitments while a few outsiders who joined the action got themselves arrested, which was not part of Shadbush's plan. On the other hand, flash mobs organized with help from persons outside the collective (but known to them) were considered very successful. Shadbush did learn from this experience and in a debriefing session on the action, members discussed the problems of working with people they didn't know, who didn't always follow through on their commitments; in future organizing, the group resolved not to undertake actions without time to build relationships and effective working groups with all of those involved (notes from November 22, 2011 meeting).

Related to the desire for maximum participation are some common rituals that take up time. In addition to the familiar introductions that are useful for allowing participants to get to know one another, many facilitators add "icebreakers" intended to add an element of fun and get people involved. These probably appeal to younger people more than older people, judging from the rolled eyes I exchanged a few times during these rituals, but they definitely add time to introductions and are sometimes done even when everyone knows one another and there is really no ice to break. While introductions and icebreakers can be related to the goals of the meeting (e.g., commenting on one's past experience), they are often completely disconnected (e.g., mentioning one's favorite vegetable). Another common practice that takes a lot of time is dividing into small groups for discussion, and then returning to the larger group to report back on the discussions. This practice could encourage more people to talk insofar as smaller groups are more comfortable for conversation, but it might also decrease the quality of the discussion if more diverse views are offered in larger groups. It seems to work best when participants join small groups based on their interests and have particular tasks to complete. When all of the groups are asked to consider the same issues, there is a lot of repetition in the report backs, and a single discussion in the larger group would likely be more fruitful. Participants differ – again possibly by generation and movement experience – in how they feel about dividing into groups. I witnessed older people on several occasions object to the plan of a young facilitator to divide into groups for discussion, as in the case of the Transition meeting described in Chapter 7 and the counter-summit on pipelines described below.

Group Cultures and Styles

Groups exhibit different styles of activism and internal cultures, which affect their efforts to devise strategies, both within groups and in coalition efforts. As Lichterman (1996, 2005) demonstrates, some groups rely on community leaders while others nourish individual expression. Rose (1997, 2000) and Obach (2004) both consider the impact of class differences in coalitions between labor and the environmental movement. Rose argues that class differences affect knowledge, and that working-class people tend to base their knowledge on their own experiences rather than on abstract analysis. Obach (2004:244) recognizes the importance of culture but suggests that class differences in the cultures of working-class and middle-class groups may be exaggerated by Rose's focus on conflicts over logging.[1] I do not have systematic data on social class backgrounds, but my impression is that more educated participants were more interested in talking about abstract issues and long-term strategies than those from working-class backgrounds, who were more likely to focus on immediate concerns and details. Whether or not differences in approach are based on class or other experiences, my observations show that group members held different views on what it meant to discuss strategy and which aspects of strategy were important. Because organizers typically did not provide much guidance on how to talk about strategy, discussions were often disjointed. Thus, a lack of strong leadership combined with different styles of organizing to make strategic planning difficult.

In some instances, the differences that surfaced in discussions about strategy seemed related to ideological variations among groups working together. For example, when one member of the Marcellus Protest CC said we should discuss "strategic messaging," a Shadbush representative replied that we were supposed to be discussing "strategy." Their exchange suggested discordant ideas about what elements of strategy an umbrella group such as MP could appropriately coordinate and what should be a prerogative of member groups. Shadbush had a more radical perspective than some other environmentalists and wanted to do its own framing of issues rather than adopt a coalition frame. When the first person argued that strategic messaging is part of strategy, the Shadbush member countered that it was impossible to agree on a "core message" and that it was fine to have multiple messages from different groups in the movement (fieldnotes, May 2, 2011). While some coalitions and campaigns adopt a common frame (e.g., Croteau and Hicks 2003; Mayer et al. 2010), this is less likely to occur without leadership that intentionally works to align the frames of diverse groups (Gerhards and Rucht 1992).

[1] There is a long history of "jobs versus the environment" discourse, which alliances between labor unions and environmentalists have worked to overcome (see Gottlieb 2005; Taylor 2016).

Within movement organizations, the differing perspectives of individuals affect strategic decision-making. This could be seen at a meeting where the CC was discussing goals, when some members talked about long-term strategic goals, but others thought in much more concrete terms. In discussing the goals for an action, for example, one member said she thought we should get a room and a big name to draw people (fieldnotes, April 11, 2012). At a meeting called to discuss strategic messaging, a participant interrupted a discussion about framing of the message to talk about the pressing need to get more literature printed and what image should be used on the front of a pamphlet (fieldnotes, July 11, 2011). Nor was she the only one to raise practical concerns in the midst of a more theoretical discussion; I frequently observed this type of (possibly class-based) division in diverse groups between those who saw the need for long-term and/or more abstract strategic planning and those who wanted to attend to practical matters. Some members did not see the need for discussions of goals and messages; they just wanted to plan actions and events. And some did not want a lot of meetings to plan actions, they just wanted to call them. There were also some disagreements about what tactics to employ, such as differences over the value of electoral politics and legislative lobbying, but these were not major disputes. Most members were open to a variety of tactics; the problem was organizing people to pursue them.

Shadbush had a different internal culture than Marcellus Protest and, as a more ideologically homogeneous group with a radical identity, enjoyed much internal accord on matters of strategy. The collective regularly spent time discussing goals, tactics, and alliances, and often had very thoughtful strategic discussions. In discussing follow-up actions to be taken after the action camp, for example, the group attempted to clarify its goals, which included influencing the media narrative, spreading direct-action tactics, and trying to mobilize participants for further action (fieldnotes, October 27, 2012). Such discussions seemed useful in getting Shadbush members to consider various options and what they could realistically do, but they tended to be inconclusive simply because it was very difficult to figure out the best strategies and sort out goals. The small group often seemed to overestimate its capacity – despite repeated warnings about this problem from several members. On the other hand, the collective did accomplish a lot; its members were very energetic and they had an extensive network of friends who helped out with various actions. A few members had part-time jobs, or were unemployed, for periods of Shadbush's life span, and were able to contribute a considerable amount of time to projects such as the action camp.

While Shadbush completed some impressive projects and engaged in numerous actions, the group also experienced some of the same problems as Marcellus Protest in devising strategies. Collective members agreed on the need to engage in strategic planning and to spend time doing so, but there never seemed to be enough time for the process. Many ideas were proposed when the group was brainstorming strategies, and of course, most of these were not followed through on. Although Shadbush members were on the same page ideologically, some were more focused on concrete tasks, such as planning a particular event, as opposed to broader strategic concerns. For example, at one of the first meetings I attended, when strategies such as working with youth or working in rural communities were proposed, some members also wanted to discuss what it meant to be a "multi-issue" group. But when time was short, the facilitator suggested that we table this "broader and more vague" discussion until there was time at a future meeting (fieldnotes, August 9, 2011). In this instance, Shadbush's practice of having a facilitator helped to move the meeting along. However, despite attempts to schedule a discussion soon, there were too many other matters that required immediate attention (fieldnotes, August 29, 2011) and the discussion ended up being postponed until a retreat in May 2012.

It seemed difficult to have discussions where everything was thoroughly hashed out and conclusions were reached, in part because of time constraints – though Shadbush was willing to have long meetings – and in part because some strategic issues were just very difficult to resolve. In a discussion of the importance and meaning of direct action, a Shadbush member commented that "direct action is meaningless without a reason to use it" and others agreed, but the meaning and goal of direct action was not clarified (fieldnotes, August 29, 2011). Despite its homogeneity, there were differences among Shadbush members with regard to their priorities and perspectives. In fact, some of the differences in ways of talking about strategy that I observed in MP also appeared in Shadbush discussions. For example, during the discussion of the meaning of direct action, one member turned to a concrete idea, volunteering that "banner hoisting" is a useful skill that he could teach to people in Pittsburgh, without discussing how this might fit into a larger strategy (fieldnotes, August 29, 2011). Some members with more organizing experience might have played a stronger role in directing discussions of group strategy, but they seemed hesitant to impose their ideas on others out of respect for participatory democracy.

The Allegheny Group (AG) of the Sierra Club (SC), like Shadbush, is a relatively homogeneous group with little internal disagreement on strategies and tactics. Whereas Shadbush had a radical identity, the AG is a more mainstream group that includes in its repertoire institutionalized tactics such as lobbying and electoral campaign work. The group receives some direction from the state and national Sierra Club and their staffs, but strategic planning is difficult to accomplish at AG Executive Committee (ExCom) meetings, as it is at other movement meetings. The AG participates in regular tactics such as tabling at events and educational forums and also works on the political campaigns of local, state, and national candidates endorsed by the Sierra Club. It has often been difficult, however, to figure out how to go beyond these familiar tactics. As in other groups, discussions of strategic objectives are frequently scattered, and it is easier to discuss concrete tasks or plans than larger strategic goals. For example, in a discussion at one meeting of what speakers to bring in, the ExCom chair asked, "what is our objective?" In response, one member talked about the need to pin cuts in transportation on state Republicans and others suggested particular people to talk to, but there was no clear discussion of objectives. After a few minutes of unfocused discussion, the chair turned to specific suggestions of who to invite to speak (fieldnotes, May 14, 2012). In another example, when discussing the upcoming climate march in New York, the AG made a short attempt at talking about the broad strategic importance of the tactic before focusing on the logistics of organizing buses and mobilizing people to attend (fieldnotes, August 11, 2014). In general, it seemed much easier to discuss concrete plans than larger goals or long-term strategies, particularly when there was no culture of discussing the latter.

Organizational Structures and Resources

As discussed in Chapter 2, the organizational structures and resources of groups affect their ability to develop leadership teams and provide guidance on the development of strategies. Effective leadership teams have meaningful discussions, gather information from a variety of constituents, and create organizational stability (Ganz 2000). Organizations that engage activists and develop their skills are likely to be more active, but many do not have the structures to facilitate recruitment and development of activists (Han 2014). As Ganz and his team found in their study of Sierra Club chapters and groups (Andrews et al. 2005), few devoted attention to leadership recruitment and training, prompting Ganz and Wageman (2008) to develop workshops on leadership teams. In accord with these findings, my observations suggest that it remains very difficult for grassroots groups to create such teams. Although they faced different challenges associated with their structures and cultures, it was difficult for all of the groups I studied to organize in ways that promoted meaningful interactions, stability, and ongoing recruitment. Moreover, discussions of leadership, and certainly efforts to train leaders, were rare in the groups I observed.

Marcellus Protest kept trying to develop forums that would allow meaningful discussion aimed at creating effective strategies, and it continued to experience difficulties based on different understandings of how to discuss strategy and a lack of strong direction from organizers, who tried hard to promote democratic participation. The special meeting of the CC called specifically to discuss strategy in August 2012 was useful in accomplishing some strategic planning. But even at that three-hour meeting, which was attended by seven regular participants (including me), we struggled to arrive at a common understanding of how to discuss strategy (fieldnotes, August 11, 2012). The facilitator got us started by asking us to comment on the positive and negative aspects of Marcellus Protest, which revealed that we valued MP's role as a hub for the region, the network developed and the groups spawned by MP, the increased consciousness about fracking, and the website. Participants worried about a "lack of structure and purpose," problems with "how to engage people," the need to mobilize more resources, and the lack of strategic direction. When the group turned to a discussion of strategic goals, it was difficult to identify goals in a way that was useful to planning campaigns. One participant said we have a clear long-term goal of stopping fracking, but we lack shorter-term goals and we haven't thought about a range of targets, such as banks. As we began to articulate a variety of goals, including regional coordination, one member tried to talk about how to organize a regional conference. When another participant suggested we should not get too "detailed" while discussing goals, the first person argued that we need to figure out how to implement goals concretely. A third participant said we needed to talk first about goals, then about specific campaigns, but also suggested that we need to talk about structure, which he saw as MP's real problem. He said, "we can choose goals, but we might keep the same structure," which would be a problem insofar as "the structure doesn't bring new people in." Although the discussion was not very systematic, the group ended up deciding on three main goals, which a participant summarized as "regional coordination, growing the group, and outreach and education," and it was at this meeting that we came up with the idea of holding informative Frack Forums to attract participants.

Even though the CC tried to develop ideas for campaigns in advance of these forums, some of the same difficulties in discussing strategic ideas persisted. The first Frack Forum, held on November 17, 2012 and attended by twenty-eight people, began with a potluck lunch, followed by a PowerPoint presentation by a CC member to update attendees on the movement before a discussion of strategic ideas. CC members divided up facilitation of the meeting, and two members led a discussion of possibilities for a new campaign, beginning with an idea that the CC had of a campaign to protect water in partnership with Marcellus Outreach Butler (MOB), which was supporting Butler County families that had lost their water. One possible target was Michael Krancer, who was appointed head of the Pennsylvania Department of Environmental Protection (DEP) after Republican Governor Tom Corbett took office in January 2011 and who had been receiving media coverage for his antienvironmental policies and clashes with the federal Environmental Protection Agency (EPA).

These suggestions sparked a wide-ranging discussion, with participants offering all kinds of ideas about targets, tactics, and means of mobilizing. At one point, a CC member expressed his frustration with the unfocused discussion by asking participants to please concentrate on the idea of planning a campaign rather than making announcements and talking about isolated actions. In an effort to incorporate the various interests of participants, three ideas were eventually identified: water/DEP; defending the Pittsburgh ban; and saving the Ambridge reservoir in Beaver County from fracking. We were going to divide into groups to discuss them further but ran out of time. Instead, the facilitator asked for volunteers to be point persons for the ideas and the CC would try to follow up on them at its next meeting (fieldnotes, November 17, 2012). As this discussion illustrated, it was extremely difficult to plan a campaign, even when the CC was trying to provide leadership to the larger group while also encouraging input from participants and working within time constraints.

Marcellus Protest did have a network of allies that could be involved in strategic planning, but the group did not have a structure with which to organize campaigns. At a Frack Forum on April 11, 2013, the CC organized a panel of activists from different groups, including MOB and South Hills Area Against Dangerous Drilling (SHAADD), to talk about their efforts and what they had learned about mobilizing participants and engaging in collective action. This was a useful discussion that focused on the importance of tactics for bringing people into the movement, particularly after an educational phase of organizing. As Joni, an experienced organizer, commented, "you have to have tasks that different people can do." Han (2014:122) further distinguishes between low-engagement groups that focus on "giving volunteers work that was as costless as possible" and high-engagement ones that worked at "creating interdependent work for volunteers that gave them some strategic autonomy, was integrated into larger campaigns, and was supported through significant training and coaching." While groups often recognized the need to engage participants, the level of engagement that Han describes was difficult for grassroots groups to achieve and maintain due to their lack of adequate structures and leadership.

There were also different views among attendees at the Frack Forum about how to continue the movement after educating people about fracking, and their comments suggested that different strategies and tactics were needed in different political and cultural contexts. People from rural areas acknowledged their need to move beyond educational forums but said that the strategy of supporting a ban that was used successfully in Pittsburgh would not work in conservative townships. An attendee from Pittsburgh suggested that, if your local council is not with you, "why not run a candidate? You can't get people to march and march and march." An organizer from MOB countered, "we haven't marched enough." There were also disagreements about whether it was worthwhile to support regulations on fracking as opposed to a total ban. But as one person commented, despite these differences, "we're all in this together" and there is room in the movement for different tactics. Another person suggested a strategic planning retreat for the "leadership" who could come up with short- and long-term goals and devise activities. A member of the CC responded that we were having a strategy workshop on June 1, 2013 – though this turned out to be a very different forum than a "leadership" retreat might have been (fieldnotes, April 11, 2013). As this and many other discussions made clear, it was very difficult to find an appropriate forum for strategic planning in a diverse movement organization that tried hard to be democratic and inclusive.

As a homogeneous collective, Shadbush had an easier time discussing strategies, but still needed appropriate forums for extended strategic discussions. The group's three overnight retreats built solidarity among members and provided opportunities for lengthy discussions of strategy. At the first retreat, held in May 2012 at the home of a member who lived in a rural area outside Pittsburgh, Shadbush set some goals that included spreading and escalating direct-action tactics, building relationships with other groups in the radical environmental movement, both regionally and nationally, and supporting local community groups such as MOB. At a number of meetings following the retreat, the group talked about these goals and how and to what extent they were meeting them (fieldnotes, November 30, 2012). At the second retreat, in February 2013, the group discussed additional goals that included expanding the national Rising Tide network in the Marcellus Shale region, building the movement in Pittsburgh, and spearheading a campaign rather than continuing to engage in one-off actions (fieldnotes, February 24, 2013). At the third retreat, in February 2014, Shadbush continued to discuss ongoing strategic plans and goals but this was probably the least successful retreat. Several members of the collective noted in interviews that, by the time of the third retreat, Shadbush had lost capacity and some of its focus. As Jessica noted, "the group wasn't able to carry forward on what we talked about in a way that we had been at the other retreats" (interview, May 4, 2015). By the end of the year, Shadbush was no longer meeting regularly and was replaced with Three Rivers Rising Tide. The retreats provided an opportunity for extended strategic discussion, and Shadbush did follow up on a number of the decisions made at them, but it was still difficult to find time to talk about everything and members found the retreat agendas very full. Even for a group as committed as Shadbush was to thinking about strategy, strategic discussions posed difficulties.

As part of a federated organization, the Sierra Club's Allegheny Group coordinates its strategies to some extent with the Pennsylvania chapter and national organization. National and state staff and volunteers provide strategy suggestions and training opportunities to the local groups. SC staff located in Pittsburgh give some strategic direction to the AG by involving volunteers in their projects and offering strategic suggestions to the group. However, only a couple of staff members have remained constant for a number of years, with some positions being cut, such as a Beyond Gas staff member. In some instances of staff positions that were cut, such as a communications staff member, AG volunteers barely knew what the staff member did. However, several other staff had strong relationships with the AG, including Rachel Martin Golman, who recruited a number of the current AG activists before becoming a regional coordinator for the Sierra Club working out of Pittsburgh. The Beyond Coal staff member, Randy Francisco, also had strong relationships with AG members, and he included volunteers in his projects whenever possible. As someone known to many other groups, he helped to link the AG to the broader movement community, but his responsibilities in the national campaign typically involved work with groups in rural areas, such as the Center for Coalfield Justice, rather than with Pittsburgh activists. Tom Hoffman, a former member of the ExCom with close ties to AG members, was hired as a staff member by the SC's Pennsylvania chapter to work on the Clean Rivers Campaign (CRC) in the Pittsburgh United alliance. He worked closely with AG members on that campaign, as well as a broader Our Water Campaign that was added in 2017 and later merged with the Clean Rivers Campaign into the Our Water, Our Rivers campaign. Tom attended some of the ExCom meetings that I observed as did Eva Resnick-Day, another staff member hired by the state chapter.

Transition Pittsburgh differed from other groups in my sample insofar as it did not attempt to devise political strategies or participate in campaigns targeting authorities. Rather, the group tried to initiate and support local projects aimed at reducing carbon footprints and creating models of sustainable living. Transition nevertheless faced some of the same dilemmas as other groups insofar as it was often difficult to design projects at its meetings and to follow through on ideas. A few active participants were interested in having a longer discussion about Transition's goals, structure, and projects. Alex Dale, a Transition member with experience in another organization with a more conventional structure, Engineers for a Sustainable World, distributed some ideas at a Transition potluck meeting about a set of leadership roles. He proposed that Transition create a steering committee with specific positions and responsibilities such as directors of communications, outreach, education, and technology (fieldnotes, January 21, 2012). Later, he helped to organize a special four-hour meeting in July 2012 to rethink Transition strategy and structure. That meeting, which Alex facilitated, was the most sustained effort at strategic planning that I witnessed in Transition Pittsburgh.

Alex had put out a call for the meeting to a list of about fifty people, but it was attended by only eight of us after we confirmed our attendance in advance. The meeting began with an icebreaker game in which two contestants at a time played a version of rock, paper, scissors and the losers had to admit to their "guilty pleasures" such as air conditioning or ice cream. Following this exercise (which seemed to me to appeal mainly to the young people in the room), we went around the circle and talked about where we saw Transition Pittsburgh going, or where we wanted it to go. This produced a wide-ranging discussion of some of the Transition-related projects taking place in Pittsburgh and of projects we would like to see happen. We talked about neighborhood projects and the potential role of Transition as a "hub" connecting various projects. Margaret described some of the existing projects in Pittsburgh but felt that the role of Transition in connecting them was uncertain. Justin said that Transition trainings were supposed to be followed up with concrete actions, but "this didn't work too well," suggesting a lack of leadership and structure in the organization. We talked about the possibility of creating a team that would hold some trainings and then follow up on them. After a long discussion, we came up with a list of "tangible goals," which included doing Transition training in neighborhoods, attending and encouraging neighborhood events, doing skills training or "re-skilling" to teach people skills that were being lost such as growing food and repairing things, holding educational nights such as films in rotating locations, and creating a speakers' bureau. After more wide-ranging discussion, we agreed on some immediate actions needed, including identifying public events to participate in and finding a person to coordinate Transition trainings and follow up on them. We also discussed current projects underway in the city that were loosely associated with Transition and how Transition could further aid these projects as well as our broader vision and goal of "trying to build resilience in Pittsburgh" (fieldnotes, July 7, 2012).

In addition to considering goals and projects, we talked about Transition Pittsburgh's structure and decided to have both a steering committee, made up of committed volunteers who could contribute 15–20 hours per month to run projects and recruit and train other volunteers, and an advisory board, comprised of people with expertise and connections who would be available to provide advice no more than five hours per month. The steering committee would consist of the directors for education, communication, archives, and outreach. Two of the positions, education and communication, were filled by Alex and Justin, while others were yet to be filled. It was not entirely clear what the next step would be, but the group had approved structural changes and discussed plans for concrete projects (fieldnotes, July 7, 2012). Although I left the meeting feeling optimistic about Transition's future, it turned out there was little follow through on the decisions made by the group. In retrospect, there were not enough people present at the planning session who were willing or able to commit the volunteer hours that Transition needed. Contrary to the Transition mantra that "whoever comes are the right people," the group really needed a larger number of energetic organizers.

STRATEGIC PLANNING SESSIONS

Recognizing the difficulties of having in-depth discussions of strategy at regular meetings, groups sometimes held extended strategic planning sessions for their members as well as strategic workshops, which were often open to members of the larger movement community. In the following sections, I discuss sessions held by the Sierra Club AG as examples of internal strategic planning efforts and then turn to a variety of strategic workshops offered by different organizations. In addition to the common problem of overly ambitious agendas and the lack of time for discussion, my observations show that the purpose of these sessions, or segments of them, is not always clear. They often include a mixture of training, educational information, assessment of tactics, and brainstorming of strategic ideas, but there is not always a connection to concrete campaigns and actions. When follow-up actions are suggested, leadership and structure may be absent, or there may be a lack of connection to the workshops. Often, the goals and impacts of strategy sessions and workshops are uncertain.

Sierra Club Strategy Sessions

The Allegheny Group held a number of special strategic planning sessions or internal workshops on Saturdays during the period in which I observed the group. One held in February 2013 was the first such session organized by the group in about five years. The previous session was considered quite successful insofar as the action teams, which supplemented issue committees, were created at that time. In 2013, there were no such changes in structure made, but the AG tried to assess its strategic priorities and to come up with new ideas for expanding the group and its campaigns. Another session was held in January 2014 followed by one in December 2014 to assess what had been accomplished in the year, followed by a session in January 2015 to further discuss the campaign ideas. Four years later, in March 2019, another strategic planning session was held. These sessions were all useful to the AG in some ways, but they also revealed some of the challenges associated with strategic planning sessions.

Finding Time for Discussion

As with the Shadbush retreats, the agendas for all of the Sierra Club sessions were very full despite their multihour formats. Ironically, one of the problems with these strategic planning sessions, which were intended to provide time to think about strategy, was that there was often not enough time for discussion. Most agendas included a mix of discussion, planning, information, and training, which competed with one another for time. Typically, there was not enough time for assessment of AG activities – even though frank and useful discussions did take place at each session. The February 2013 session, which ran five hours including lunch, exemplified these problems. It was attended by twenty-three people, mostly long-time group members, and cofacilitated by a national staff member who worked out of the Pittsburgh field office and a volunteer. The format was a mixture of information provided by leaders and brainstorming of group priorities and plans, though time was short for discussions throughout the day. Some parts of the program did allow for discussion, including "breakout groups" that were each assigned a topic (advancing the Sierra Club mission; providing service to members; building local resources; or effective governance). One of the facilitators stressed that "this is not a training" and that the groups should discuss the strengths and weaknesses of the AG in these areas, which were detailed in a handout. In the group I participated in, the discussion was lively and interesting, but cut off by a 20-minute time limit followed by 20 minutes for report back.

Some portions of the strategy sessions demonstrated the value of time set aside solely for discussion. In the January 2014 strategy session, the first meeting in the morning was devoted to discussion among issue chairs on how to carry out goals. Participants included several members of the AG from areas outside Pittsburgh and was, by all accounts, a very useful discussion that identified opportunities for involving activists and organizations from counties surrounding Pittsburgh in Sierra Club activities. More typically, useful discussions had to be cut short to accommodate other parts of the agenda. In December 2014, the AG held another 5-hour strategic planning session that was intended to assess the group's efforts during the year and think about what might need to be changed, setting priorities for 2015. The ExCom chair made it clear that the goal was to assess everything, including the action teams, various activities of the group, and its campaigns and coalition work. While the agenda was overly ambitious, the session produced frank discussion about the structure and work of the AG. After an introductory overview of the action teams and their history, we divided into three groups to discuss the current action teams on global warming, endangered species, and public lands before reporting back to the large group. The time allotted for both small and large group discussions was inadequate, leading one participant to complain in a good-natured way that "arbitrary time lines interfere with the creative process!" The ExCom chair responded in kind, "yes, but it probably won't change!" Although this was a light-hearted exchange, it pointed to a tension between allowing time for brainstorming and keeping on schedule. Shadbush probably would have extended the time so as not to cut off the discussion and then gone over the time for the meeting, which might have been more satisfying to participants who value participation highly. But the Sierra Club was more predictable in providing a structure to meetings and events, which had the advantage of allowing busy members to schedule their time accordingly and may be one reason the Club has survived for so long.

Connecting Sessions to Group Activities

In addition to the frequent lack of adequate time for discussion, some of the training and information content, often provided by staff members or volunteers who were not core activists in the group, was not directly connected to the group's actual work. At the February 2013 workshop, the chair of the Pennsylvania chapter of the Sierra Club came to talk about the characteristics of successful chapters and passed out an informational sheet on this. She explained how the chapter can support groups with expertise, ideas, and other resources, but after her presentation, there was no discussion of how the AG is using chapter resources and how specifically the group might work more closely with the chapter on its projects. Later in the day, there was a "training" component to the strategy session (labeled as such in the agenda), which included a discussion of the components of campaign planning by the volunteer facilitator (using the Midwest Academy model[2]) and a "communication and

[2] The Midwest Academy is a training school for organizers founded in 1973. Its widely used training manual and strategy chart direct organizers to think about their goals, organization and resources, supporters, allies and opponents, targets, and tactics (see Bobo et al. 2010).

messaging training" by a staff member. This information was useful but might have been more meaningful if it was connected to planning an actual AG campaign. In the case of the messaging training, the presentation pointed to resources on what was then the national organization's "Clubhouse" website (currently "Campfire"), such as instructions for doing a "message box" to aid in framing issues. I was surprised to learn that group members were previously unaware of this resource. Thus, the strategy session provided opportunities for members to learn more about the Sierra Club, but also demonstrated the lack of sufficient coordination between levels of the organization.

At the January 2014 strategy session, there was a workshop in the afternoon by Tom Hoffman, who had many years of organizing experience in unions and environmental groups. The workshop had the goal of teaching participants "to think as organizers" and engage in effective collective action, which involved breaking into two groups and discussing the Clean Rivers Campaign, which was currently underway in Pittsburgh, and what strategies it might pursue to be successful. In the discussion following the smaller group discussions, we talked about strategies for gaining power and some successful past Sierra Club campaigns. Next, we did another exercise related to the American Legislative Exchange Council (ALEC) and its various proposals to remove environmental protections and safety regulations. In the evaluation of the workshop held at the end, participants expressed positive sentiments about these exercises but also wanted to talk more about specific tactics being used by the AG and its action teams and particular campaigns that might be waged (fieldnotes, January12, 2014).

Despite time limitations, the AG made some assessments of the strengths and weaknesses of its action teams and other activities at its strategy sessions. At the December 2014 session, the Endangered Species Action Team (ESAT) reported as its strengths regular meetings and socializing among core members and regular activities such as tabling and doing educational presentations at schools. In terms of weaknesses, the team felt that its focus and activities were somewhat narrow and not well integrated with the rest of the AG's work. The Global Warming Action Team (GWAT) was less upbeat, feeling that their core of about five regular activists was too small and that many of their activities, such as tabling and making phone calls on activist nights, were only indirectly related to global warming. During the small group discussion, the team was just about to discuss what role it might play in a new Pittsburgh Climate Action Network, spearheaded by Sierra Club member Peter Wray, who was present, when we ran out of time and had to return to the larger group (fieldnotes, December 12, 2014).

Goals and Outcomes

Like the problems encountered by Marcellus Protest, there was a fair amount of uncertainty when it came to discussing goals. GWAT admitted to having only long-range goals (limiting climate change) rather than short-range ones, and

ESAT felt it needed to broaden its concerns. At the December 2014 strategy session, however, ESAT members did not address the question of what impact their activities might be having, whereas at an ExCom meeting earlier that year, a team member wondered if they were "preaching to the choir" with activities such as presentations at the Environmental Charter School, where the children typically have parents who are already concerned about environmental issues (fieldnotes, May 14, 2014). We discussed how we might measure success, but also the problem that we tried to do that at a previous strategy session where we came up with goals, and the action teams and issue chairs were unable to show how they were achieving these. One of the problems was that most issue chairs lacked real teams or active committees and simply did not have the capacity to pursue goals on their own. The action teams were also small, and needed more activities to expand themselves, but had difficulty engaging in more activities without more members. Thus, resource limitations and the lack of enough members working on active committees lessened the impact of strategic planning. The AG needed to build "campaign-based teams" of activists with "a clear set of goals they were trying to accomplish, and strategies they were using to reach those goals" (Han 2014:105).

After discussing the action committees at its December 2014 session, the AG discussed some of their activities: tabling, participating in city recycling events, activist nights, and social activities (including the holiday party, summer picnic, and St. Patrick's Day parade). We again divided into small groups, with each assigned an activity to discuss, and then reported back to the larger group. This process created some difficulties, as one group (recycling) had no one in it who actually volunteered for that activity, and at least one group (social activities) was very small, with only two other persons besides me. But the group came to some conclusions: Tabling was seen as valuable for providing exposure to the group, collecting names on petitions, and making contact with other groups that sponsored events. Volunteering at recycling events (such as collections of hard-to-recycle items) started when the AG was trying to establish a relationship with the city and make officials aware of the Sierra Club; there were mixed views as to its current importance and whether or not a campaign to expand recycling might be worthwhile. Activist nights were poorly attended, with only four regular participants, and consisted largely of making phone calls to help a staff member with his projects. On the positive side, the group succeeded in turning out members for many events. It was decided that the format needed to be expanded to include more activities besides phone banking and that the volunteers needed to redefine their roles with the staff member. Regarding the social activities, it was thought that both the picnic and holiday party provided opportunities for extended conversations with others in the environmental community who were invited to the events. Participation in the St. Patrick's Day parade, which was originally thought to have educational value, had become burdensome and it was decided that it was not worth the time to continue.

In addition to assessing these regular activities, the AG talked about three campaigns that group members were involved in: Protect Our Parks, the CRC, and the emerging Pittsburgh Climate Action Now group. Originally, the agenda called for small group discussions but to speed things up, the organizers decided to dispense with these and just discuss the campaigns in the larger group. A participant from each gave a summary of what was happening and the role of Sierra Club members. In the case of POP, it was noted that a number of AG members were going to county council meetings to testify. Many Sierra Club members were also showing up to CRC events, which were pressuring the Allegheny County Sanitary Authority (ALCOSAN) to pursue green solutions to deal with Pittsburgh's storm water runoff from sewers into the rivers.[3] The campaign was having some success and Tom argued this was due in part to the breadth of the coalition, which included labor unions and faith-based groups as well as environmentalists. Peter added that staff are also important to its success, in recognition of the difficulty of mounting all-volunteer campaigns. Pittsburgh Climate Action Now was still at the stage of discussion among individuals, including a number of Sierra Club members.

Subsequent strategy sessions also revealed how difficult it was to pursue goals without an adequate number of highly engaged volunteers. The February 2013 session produced a set of goals, which were later summarized as improving communications with AG members; developing a larger and more diverse membership base; increasing fundraising revenues; and electing environmentalists to public office. In January 2015, the AG held a follow-up strategic planning session, scheduled for 4.5 hours. At this meeting, we first reviewed and discussed the goals that had been set at the 2013 strategic planning session in the group as a whole (rather than smaller groups). The discussion was wide-ranging and not entirely focused on the four goals. A number of participants stressed the importance of coming up with campaigns that would involve supporters and accomplish environmental objectives. But in thinking about previous campaign efforts, the group recognized the dilemma of needing more volunteers to get campaigns going before the activities would in turn bring in more supporters. Working in alliances on campaigns that were already underway was one way to get around this problem, though such efforts did not necessarily build the group.

The AG had decided to try to restructure the action teams, having one team instead of several to work on multiple campaigns. Three local campaigns that

[3] ALCOSAN was under order by the EPA to end illegal discharges of sewage into the rivers, but originally came up with a "gray" solution to the storm water runoff problem. The CRC pressured ALCOSAN to adopt a "green infrastructure" plan instead, employing wetlands, green roofs, and other solutions to absorb storm water before it caused overflowing of sewers. The CRC won a victory in 2017 when the EPA allowed ALCOSAN five years to show that green solutions could work, and in 2018, two projects were funded by the city of Pittsburgh to achieve this goal (www.cleanriverscampaign.org).

the AG would work on were identified, including an endangered species project and two coalitions, POP and the CRC. Climate change was seen as an overriding concern for all of the campaigns and the new combined team was later called the Endangered Planet team. We then divided into small groups based on these three campaigns to discuss questions that were provided before returning to the larger group for report backs. The group discussed what the Sierra Club could best contribute to POP and the CRC; in the case of the CRC, having a staff member directing the effort made this easier and many AG members attended events organized by Tom Hoffman. The endangered species campaign was going to focus on bees, butterflies, and bats, which was very appealing to many members, but tactics still had to be worked out.[4] As usual, time was short at the end of the strategy session and we never got to an evaluation of it, but participants seemed pleased that there had been some good discussion and that the group would be focused on several campaigns.

In November 2016, shortly after Donald Trump's election to the presidency, the AG took advantage of another opportunity for strategic planning when an intern supervised by Tom Hoffman led the group in a SWOT (strengths, weaknesses, opportunities, and threats) exercise. Time was set aside for this at an ExCom meeting and we were asked to discuss what went right and wrong in the past year and what we would like to accomplish in the next year. This started out a little awkwardly because we had to think about strengths and weaknesses within a time frame rather than more broadly, but it ended up being a good discussion. In addition to specific ideas for actions and campaigns, we talked about using the threat from Trump to build the membership and defending environmental protections in the new political climate. We also talked about the recent recruitment of four relatively young new board members as an opportunity to bring in new ideas and energy. Tom joked that now only half of the ExCom was on Medicare! This was funny but also pointed to both the important role of retired people with time to contribute in the Sierra Club and the real problem that many environmental groups have had in attracting young people. We discussed the problem of a lack of enough ongoing activities to involve new activists and, once again, some good ideas were proposed, such as an urban outings program. We ended up with an exercise created by the national Sierra Club that involved placing dots on a Chapter Capacity Wheel (fieldnotes, November 15, 2016). In January 2017, the intern, who also conducted individual interviews with a number of leaders and members of the AG, distributed a report summarizing his assessment of strengths and opportunities as well as weaknesses and threats facing the group. We discussed the report at the ExCom meeting and the need for more

[4] As it turned out the campaign did not get off the ground at this point, but the AG later developed an educational program about bees, butterflies, and bats as part of a project in the neighborhood of Wilkinsburg begun in 2017, which involved working with residents to create pollinator and vegetable gardens.

campaigns and activities to build the organization. Tom suggested having another strategic planning session, but Barb said the problem with the previous ones is that nothing really came out of them (fieldnotes, January 9, 2017). This exchange suggested that the AG committees and action groups lacked adequate numbers of participants and leadership, and that the AG might have needed other types of structures such as a recruitment committee. The national and state Sierra Club organizations were creating plans and adding staff as part of the Trump resistance, which sounded promising as a way to expand the group and develop campaigns.

The AG did eventually hold another strategic planning session in March 2019, which was planned and facilitated by a state Sierra Club staff member and a committee from the AG ExCom. It was an upbeat meeting of both long-term members and new participants who had joined the group and the ExCom, including some young people. The group discussed what had been accomplished in 2018 and set priorities for 2019. The focus was on concrete actions that could be taken and specific plans, which seemed to energize the group (fieldnotes, March 16, 2019). But the question remained as to whether the AG had the structures and leadership needed to recruit more activists and organize ongoing campaigns.

Strategic Training Workshops

In addition to largely internal meetings such as the Sierra Club sessions described above, grassroots environmental groups held strategic training workshops of different types for the larger movement community. In some instances, activists had specific goals in mind when they planned the workshops, while in other cases, the intent seemed to be general education. Anticipated outcomes did not always occur, though workshops and other forums might have had favorable consequences such as inspiring individuals, familiarizing them with approaches to strategic planning, and training them in particular skills. They may have helped to build the movement – or they might have had little impact. In the following discussion, I examine the content and outcomes of the workshops and forums that I observed. Although most of the workshops were of interest to attendees, the following detailed examples show that their impact on movement activities was often unclear.

On December 4, 2010, well before I began attending Shadbush meetings, I went to a two-and-a-half-hour workshop on corporate research given by Patrick, a Shadbush member and experienced organizer. It was held at the Sierra Club office and initiated by another Shadbush member who, at the time, was working for the Sierra Club as their student organizer and had permission to use the office space. The workshop attracted around twenty-five people, including a number of participants in the recent MP demonstration, who were eager to learn new ways to expand their opposition to fracking. Patrick explained that he had first developed the workshop as part of the tactical training initiative of the Pittsburgh Organizing Group (an anarchist group no longer in existence), which also included other workshops, including one on direct action. At this workshop, Patrick explained why it might be useful to target corporations as part of a larger strategy and how research on corporate operations, such as gas drilling companies, might help the movement to gain leverage. He also provided guidance on how to access information about corporations through various sources such as government and shareholder reports. Attendees at the workshop were very interested in pursuing this strategy and they asked about specific tactics such as boycotts that might come out of such research. Patrick explained that boycotts are difficult to implement and that "boycotts are a tactic, not a strategy" whereas he was talking about a strategy such as targeting institutional landowners leasing their property for fracking.

Despite the strong interest in corporate research among workshop attendees, no one ended up organizing a group that would follow up on the training. Patrick suggested that Marcellus Protest might want to get together a research team, saying he could be a resource, but was not offering to lead such a group (fieldnotes, December 4, 2010). He had been asked to do the workshop by the Shadbush member who worked for the Sierra Club, but was not prepared to organize a campaign, which would have been a major time commitment (interview, April 12, 2015). Later, Patrick did some research himself on the investors behind the Marcellus Shale boom, which was published in a report on the Shadbush website and in the 2013 *Shalefields Grassroots Reporter*. The failure to create an ongoing committee following the workshop illustrates the need for leadership and structure in a grassroots movement; MP did not have a structure capable of supporting ongoing campaigns or standing committees unless individuals took the initiative to organize them.

Strategy workshops require a great deal of work to prepare, and effort needs to be put into attracting an intended audience as well as creating useful content. On October 22, 2011, Shadbush conducted a six-hour workshop on direct action, which was attended by thirteen people, seven of whom (including me) were members of Shadbush. Although the attendance was disappointing, the workshop was well received. It included segments prepared and presented by the six Shadbush members (all except me) on the history of the environmental justice movement, the meaning of direct action, the importance of goals, strategies and tactics, media coverage, logistics, and civil disobedience and the risk of arrest. Shadbush envisioned this workshop as a precursor to planning the demonstration at the DUG East conference and announced a follow-up meeting to be held at the Occupy Pittsburgh camp the next week to plan this action. However, most of those who became involved in the DUG East action did not attend the workshop, which Shadbush hoped would lay the groundwork for working together on the demonstration. In this case, it would have been useful to recruit more of the same people for both the workshop and the action planning, but as a small collective, Shadbush had limited capacity to do so.

While connections between workshops and subsequent collective actions are desirable, it takes a lot of planning and work to coordinate strategic training and action. Following the February 2013 retreat at which Shadbush decided to try to spearhead a campaign, the group held a workshop on planning grassroots strategic campaigns. The workshop, held at the Sierra Club office, was attended by twenty people of varying ages and diverse movement backgrounds. Attendees were highly engaged in the workshop, which seemed well received; it was scheduled for two hours but ran overtime by half an hour. It was facilitated by Ben and Patrick, who explained that the goal of the workshop was to show how to do a grassroots campaign by selecting an issue, tactics, and goals. They gave some background on Shadbush and explained that this workshop was connected to the group's goal of generating a campaign, but that there would not be a lot of time to brainstorm about specific strategies. They said the workshop would provide a general overview for attendees. Much of the workshop consisted of participatory discussion and some exercises and partner discussion. We talked about the meaning of terms such as campaign, strategy, target and tactic, and distinctions between long-, medium-, and short-term goals. We discussed how to set campaign goals and pick targets. Discussion at the workshop differed from the strategy discussion that often occurs at organizational meetings in that it was more focused, but as with many other strategic discussions, there was insufficient time to discuss everything because so much was crammed into the workshop. Although Shadbush saw the workshop as a first step in developing a campaign, it also had an educational goal for other groups. And, other than the Shadbush and MP members who attended, none of the others who came to the workshop participated in the next session called to devise a campaign, which was held several months later.

Workshops sponsored by Marcellus Protest also had uncertain goals and impacts. Like Shadbush, MP was also looking to jumpstart the movement by creating a new campaign. It happened that the group was contacted by a staff member for Food and Water Watch (FWW) who offered to give a free strategy session for MP, which decided to accept the offer. The workshop was held on June 1, 2013 and attended by about twenty-five people, some of whom were active in organizations, including MP members who hoped that the session would help them to develop a new campaign. The young staff member from FWW did a presentation, using materials from the Midwest Academy, to explain the elements of creating a strategy. He talked about direct action organizing and the steps involved in creating an effective campaign, including identifying a target and building power, and criteria for choosing issues. Then he had us do an exercise using the Midwest Academy strategy chart, which he explained before we broke into smaller groups. We were given a sheet describing a water privatization fight and instructed to work out a strategy using the chart, which we worked on for an extended length of time before reporting back to the larger group with 15 minutes left in the allotted time. Participants who had been unfamiliar with the Midwest Academy approach commented that they found it useful, and the FWW staffer offered to be a resource to us in developing a campaign related to fracking; he said he picked an example unrelated to fracking so we could focus on skills in the workshop. In conversations afterwards, however, participants expressed disappointment that we had not worked on coming up with a new strategic campaign to oppose fracking. While the workshop was intended to develop our skills in devising successful strategies, it is not clear what impact it had on individual participants or their organizations.

Workshops aimed at teaching particular strategies may be more influential, though their impact is also uncertain. In January 2011, the Outreach committee of MP sponsored such a workshop, which was a presentation by Ben Price of the Community Environmental Legal Defense Fund (CELDF) on the community rights approach that was used successfully in the Pittsburgh ban on fracking. The workshop, which lasted over six hours, was held at a location about 20 miles outside of Pittsburgh and attended by eighteen people in addition to the presenter. Ben explained that the aim of the workshop was to provide a step-by-step guide for going into the community and getting a ban on drilling passed, as was done in Pittsburgh. He gave a lot of legal and historical background, and explained how the community rights approach, which asserts the right of local communities to ban practices such as fracking or factory farming, differs from other approaches such as the use of zoning and other regulations. He explained that the goal of organizing was to stop fracking by asserting community rights to self-government and a healthy environment. He provided a sample ordinance and explained the elements of it in detail. And he gave a detailed description of how to go about drafting an ordinance and promoting it in a local community. He said that CELDF could provide advice but not extensive legal services to communities. Participants in the workshop were very interested in the approach and had a lot of questions. While some hoped to pursue this strategy, there were concerns that it was not feasible in many conservative communities and that municipal governments were often fearful of being sued if they passed a ban on fracking. The workshop was highly focused and provided extensive information, and CELDF later held additional workshops in Pittsburgh sponsored by various groups. While it is uncertain what impact this particular workshop had, the CELDF approach did become familiar to many in the movement community, and community rights bans on fracking were adopted in a number of other municipalities besides Pittsburgh. Terri Supowitz, who attended the CELDF workshop, later spearheaded a successful effort in Wilkinsburg to get a ban on fracking passed, which included having Ben Price talk to the borough council (interview, April 23, 2012).

DEVISING MOVEMENT-BUILDING CAMPAIGNS

Movements participants are generally quite interested in strategic planning, but they are often frustrated by strategy sessions and workshops that are not directly related to actions and campaigns and by the difficulty in following up on strategic planning. Nevertheless, organizers keep trying to generate actions and campaigns because they understand their role in building the movement, countering opponents, and achieving goals. In the following sections, I provide extended descriptions of two efforts at creating campaigns that would build the movement against fracking: an attempt to generate actions around pipelines and the creation of the Protect Our Parks campaign. Organizers of a counter-summit on pipelines hoped to generate multiple actions by groups outside Pittsburgh leading up to the event and they hoped that the educational event would kickstart a larger campaign around pipelines. But organizers had little control over the decentralized movement and efforts to be inclusive with regard to agenda planning made it difficult to control the event. In the case of POP, the campaign was longer lived, and core organizers worked together well, but the coalition faced strong opposition and the campaign ultimately ceased its activities.

Planning a Counter-Summit on Pipelines

Some of the impetus for particular actions came from external threats and industry activities. Concerns about the expanding network of pipelines and other infrastructure, including compressor stations, were increasing as residents became aware of the possibility of losing land to eminent domain, giving companies the right to cut through private and public lands, and as the media covered dangers such as exploding pipelines (Lord 2013; Seelye 2011). During a Marcellus Protest CC meeting, David searched online for industry events and learned that industry executives would be holding their third annual Marcellus Midstream Conference in downtown Pittsburgh from March 19–21, 2012 to focus on infrastructure. The CC decided that the event would be an opportunity for self-education on the issue of pipelines and infrastructure and also a way to revitalize the movement. Many different ideas were discussed at the meeting, but the focus on planning for a particular event seemed to help the group develop a strategic plan. Mel argued that he did not want to do "just a protest" but something useful for "firing people back up, getting people involved again" and "not just a one-shot affair." Ben questioned whether we wanted to respond to an industry event, suggesting that we "think more about movement building, and think about organizing not just in response to industry events." But others saw the industry event as an opportunity to draw attention to the impacts of oil and gas drilling infrastructure. We discussed many possible tactics, including civil disobedience, but also the difficulties in staging a large protest with our limited capacity. For a while, it seemed like it would be difficult to come to a consensus about what the action would entail, but we eventually agreed on a small demonstration followed by a dinner and educational event billed as a "counter-summit." We wanted to involve people in surrounding communities and to avoid a one-off event, so we talked about having events in advance of the summit such as film showings with our allies and getting local community groups such as MOB and SHAADD to do events leading up to what was conceived as a "mini counter-summit" (fieldnotes, January 11, 2012).

The CC planned an agenda for the January 28, 2012 GA, which lasted three-and-a-half hours. Discussion of the plan for the pipelines event was discussed following a potluck lunch, short videos on fracking by a local film maker, and a presentation on strategic messaging. Although the meeting agenda was packed as usual, the GA had a good turnout of close to fifty people, and the strategy discussion gave the meeting a useful focus. CC members facilitated the meeting and explained their idea of events held by local organizers from the region in their communities in the weeks leading up to the counter-summit, at which participants from different communities would come together for a small demonstration and counter-summit on pipelines. The counter-summit would be an educational event with a panel of speakers knowledgeable about the impacts of pipelines being constructed for the gas industry in Pennsylvania. Having previously discussed the strategy at its own meeting following the CC meeting, Shadbush proposed that there also be a strategy planning session at the counter-summit. While one MP organizer expressed confusion at this idea, a Shadbush member argued that we should not let people leave the event without direction and a sense that the movement has a roadmap (fieldnotes, January 28, 2012).

As this exchange illustrates, the ability to think through and discuss strategic ideas is complicated by dynamics in alliances as well as within movement organizations. Despite the misgivings of some MP organizers that the strategy planning would take too much time and be out of place at the event, there was little debate at the general assembly. MP agreed to the Shadbush proposal mainly because organizers did not want to disappoint the young Shadbush activists who provided so much energy to the movement. As it turned out, there was simply not enough time at the counter-summit to hear about pipelines from knowledgeable speakers and to plan movement strategy. Questions for a speaker who had traveled from out of town to be at the counter-summit had to be cut short, and the attempt to do both detracted from the focus on pipelines – an issue that has since become increasingly important to the anti-fracking movement. Thus, the lack of adequate discussion of a strategic planning idea in meetings could have important consequences. While the homogeneous Shadbush collective might have come to the same decision with further discussion at their own meeting, a wider range of views might have been presented had there been more debate at the general assembly, and a different decision might have been reached.

In general, the GA felt energetic, but the presentation of ideas for the strategy by the cofacilitators provoked numerous questions as participants tried to figure out what the event would look like. They explained that we were still "brainstorming" and that we would be dividing into groups to talk about various elements of the event, including direct action, education, outreach, and messaging. This created momentary confusion and one participant asked, "what are we supposed to do in the groups?" Some of the groups had a lot of people, while a few were tiny, such as the direct-action group that I joined, which had only two other participants besides me. After the group discussions, there were report-backs to the larger group and committees were formed, each with a designated contact person. Although the process was scattered and at times confused, many people contributed a lot of different ideas as we discussed the event and most participants seemed enthusiastic at the end of the GA. When asked if we needed to have another large meeting, in addition to committee meetings, there was strong sentiment in favor of another GA, but only eleven of us showed up at that meeting, which was held on March 3, 2012.

In addition to the general assemblies, I attended two planning meetings of the education committee for the pipelines counter-summit, where we talked further about the goals of the educational and strategy parts of the event. We discussed the need to explain the impacts of pipelines in the state, such as how trails in some state parks have been taken over by pipelines, and to discuss the range of tactics needed, including civil disobedience and mass actions. The counter-summit was seen as part of a movement-building strategy; as one participant argued, its goal was to "get people to feel more connected, to have a sense of power to work in concert, to get people ready for this, to think about collective power." Another added that if we do enough "education, sharing of information and confidence building," then some of the people will be "capable of teaching others; we can supply them with tools" (fieldnotes, February 7, 2012). The committee also worked out a five-hour schedule for the event, to be held at a downtown church, which included a sign-making period, a march to the convention center where the industry was meeting and back to the church, dinner, a panel on pipelines, and a strategy session. Committee members volunteered for numerous logistical tasks such as organizing food, getting literature printed and distributed, bringing materials for sign making, lining up Pittsburgh's radical marching band for the demonstration, and contacting allies (fieldnotes, February 25, 2012). At the March 3 meeting of the larger group, we got reports from the committees and talked about tasks that still needed doing in preparation for the event. We also engaged in a fair amount of off-agenda discussion about issues related to fracking and how to frame issues – the type of important discussion that sometimes happens when a meeting is smaller or shorter than expected.

The demonstration and counter-summit on March 20, 2012 was well attended by about seventy-five people (who came and went during the event, making it difficult to get a precise count). The marching band of about six musicians made the march feel even bigger, and there were some shouted interactions with industry convention attendees that, along with speakers at a brief rally, roused our emotions. After dinner at the church, cooked and served by volunteers, we were running about 20 minutes behind schedule when two speakers gave very interesting and informative talks on pipelines and infrastructure development in Pennsylvania. There were a lot of questions for the speakers, which unfortunately had to be cut short to start the strategy planning part of the event. The Shadbush member leading the strategy discussion had us divide into four groups for a SWOT analysis of how we can stop pipelines. Each group got a large sheet of paper with four boxes to fill in, and we had ten minutes for the exercise before returning to the larger group to report on our discussions in the ten minutes remaining. Not surprisingly, there was insufficient time for strategic planning at the counter-summit and no cohesive strategy was formulated there.

Brainstorming and discussion at strategic planning sessions sometimes take a backseat to "exercises," a practice that takes time and is not universally appreciated. When the young Shadbush organizer announced that we would be dividing into groups for the strategy exercise, the seasoned activist sitting next to me whispered in disbelief, "we have to divide into groups? We only have a half an hour. I thought we were going to have an open discussion." When she heard we were going to do a SWOT exercise, she commented to me sarcastically, "how original" (fieldnotes, March 20, 2012). While some exercises are useful in training activists and stimulating discussion, they often displace open discussion and take longer than a free style discussion might. In the case of the pipelines event, it would likely have been more useful to use the short time left to simply throw out ideas about actions that might be taken on the issue – though this would not necessarily produce any subsequent actions.

As it turned out, MP's pipeline event did not stimulate an ongoing campaign, though many actions focused on pipelines have since been undertaken by Pennsylvania activists and those who attended the counter-summit may have benefitted from its educational content. In 2016, some Pittsburgh activists traveled to the Standing Rock Sioux Reservation to support the Native American movement to stop the Dakota Access Pipeline (Potter 2017). In Pennsylvania, activists became involved in campaigns against a number of pipelines, including Sunoco's Mariner East II pipeline, which threatened drinking water in southeastern Pennsylvania (Legere 2017b). While the MP counter-summit was an educational event that did not have a specific target, subsequent organizing around pipelines and drilling that impacted particular communities did spark movement campaigns. In contrast to the MP pipelines effort, the Protect Our Parks coalition launched a three-year campaign with multiple actions and tactics.

Brainstorming a Campaign

While many strategic planning sessions and workshops had uncertain outcomes, Shadbush's decision to try to initiate a campaign was successful in creating a new coalition, Protect Our Parks. After its strategic planning workshop on March 30, 2013, Shadbush invited members of MP, MOB, SHAADD, and other organizations to a brainstorming session for coming up with a new campaign, which was held on June 2, 2013 and attended by seventeen people. Shadbush had prepared an agenda for the meeting and two members of the collective cofacilitated it. In contrast to Transition's approach of "whoever comes are the right people," Shadbush members thought carefully about who it was important to have present, though they also included a few people who heard about the meeting and asked to be invited. Depending on who was present, different ideas and strategies would likely be proposed based on the experiences of the participants. The facilitators started out the meeting with a (mercifully) short icebreaker exercise. Then a Shadbush member explained at the start of the meeting that the collective had been organizing "one-off actions" such as the ones at Maggie's Farm and the DEP, but it seemed logical to go from that to organizing a more sustained campaign, where we pick a target we want to go after and articulate goals. To do this, Shadbush wanted to team up with allies in the community to choose and carry out a campaign that might give the movement a badly needed victory.

We began with a wide-ranging, open discussion of a large number of issues and targets related to fracking such as seismic testing, pipelines, and threats to water supplies, state forests and parks, schools, and college campuses. After a period of discussion, one of the facilitators then asked for point persons who would lead smaller group discussions and online research on particular issues to come up with campaign ideas. Although there were numerous ideas and it was difficult to organize them, we ended up dividing into five or six groups, and some individuals started doing research on their computers.[5] When we returned to the larger group after about 30 minutes, the facilitator suggested that we could throw out new ideas as well as try to whittle down the suggestions to a manageable number of ideas. However, it was difficult to both generate creative ideas and pare down the options in a short session, suggesting a tension between brainstorming and decision-making. We began focusing on suggestions from the small groups, which one of the facilitators summarized in six entries on a flip chart. We were then given two dots to place on the issues that interested us most, followed by a decision, in the interest of time, to discuss the pros and cons of only the top four ideas. Our discussion of the four issues included targets, goals, and frames; we were instructed not to discuss specific tactics yet. One problem considered was that it would be difficult to mobilize Pittsburgh residents for a campaign targeted outside the city. After much discussion, we decided to adopt a "public lands" theme that various campaigns might fit into and that we could develop in terms of messaging. This focus reflected some of the major concerns of activists represented at the meeting about the impacts of fracking in Pennsylvania and their sense that this focus could generate public support. Patrick suggested that the next step was to have

[5] Participants had been told ahead of time to bring laptops if possible and prepare to research issues at the meeting. While this suggested a computer savvy, middle-class group, it was not necessary to have a laptop because most participants worked in groups.

another longer meeting to develop the message and that we also needed to do more specific research on cases, invite others who are working on public lands issues to attend, and to think about organizational structure and come up with activities. We then made a list of various tasks to do before the next meeting, such as researching issues and inviting particular people. We decided to have two additional three-hour meetings in the next month rather than one longer meeting after Mel argued that three hours was long enough and others agreed. There were some mild disagreements about what additional people, from what organizations, should be invited, and one person expressed concern about having a lot of different people at the next meeting.

At the second meeting, held on July 7, 2013, a somewhat different set of people did come, with about six out of seventeen newcomers. These were people in the movement community who were invited to come because of their involvement with public lands issues, including members of a Beaver County group fighting to preserve the Ambridge Reservoir and the Public Lands issue chair from the Sierra Club. (As anti-fracking activists who were known to others at the meeting, the newcomers posed no problems.) Two different members of Shadbush cofacilitated the meeting, and one of them explained for the newcomers that Shadbush decided to host some meetings to come up with a longer-term campaign and that it was still unclear what direction the group was going in, but we are interested in public land issues. After asking us to introduce ourselves and say what summer fruit or vegetable we liked best (the icebreaker), one of the facilitators recapped the last meeting. Then we were instructed to complete a writing exercise about what we hoped to achieve in fighting fracking on public lands; when this created confusion, one facilitator explained that it was a way of getting at our goals for a campaign. As we shared our goals with the group after writing them down, they were recorded on a flip chart. We were then asked to talk about the issues we identified last time, plus a new one that came up since the last meeting, before talking about specific campaign ideas.

The new issue was the potential for fracking in the Allegheny County parks; since the last meeting, the Allegheny County executive, Rich Fitzgerald, had announced his intention to lease Deer Lakes Park for shale gas drilling under the park (*Pittsburgh Post-Gazette*, June 13, 2013). Earlier in 2013, activists had tried unsuccessfully to convince the county council not to lease airport land for fracking, but it seemed possible to flip some of those votes when it came to the parks. There were also perceptions that Fitzgerald would make a good target because he was easy to provoke and because some county councilors were reportedly dissatisfied with his leadership, which might present a political opportunity. When we did a "straw poll" to see which issues participants were most interested in working on (with multiple votes permitted), the county parks issue was the clear winner, though there was also enthusiasm for working on other issues related to public lands. We discussed extensively both the activities of the Beaver County group around the Ambridge Reservoir and work being done by the Center for Coalfield Justice and others to protect state forests and parks and how a coalition might contribute to these campaigns. Some participants wanted to support multiple campaigns and draw connections among them as part of a push to protect public lands. However, the county parks issue was immediately appealing as a current threat close to home for Pittsburgh residents, and it was decided that the next meeting would focus on developing that campaign. Ultimately, it seemed too difficult to pursue and connect multiple campaigns around the public lands theme.

At the meeting to develop the campaign a week later, a smaller group of ten people discussed strategy for the parks campaign, including targets, tactics, messaging, allies, and outreach. While Rich Fitzgerald was seen as an attractive target, county councilors were seen as potential allies who should be lobbied. The message would be that these are "our parks" – a public good that needs to be preserved. Regarding tactics, the group decided to start mobilizing people to attend county council meetings to talk about the issue; even if an issue was not on the agenda, residents could sign up to talk for three minutes during the discussion period. Other immediate tactics included getting people to write letters, call and visit their councilors, asking environmental and other groups to sign on to a letter opposing leasing of the parks, and collecting signatures on petitions at tabling events. Several people volunteered to work on creating a piece of literature to be passed out. More participants would be recruited at the next Frack Forum, which was already scheduled for the following week, and the first open meeting of the new campaign coalition was scheduled for August 3, 2013. At the Frack Forum, there was quite a bit of enthusiasm for the campaign, and a number of people who learned about it there came to the August 3 meeting.

That meeting attracted twenty-one people, including some newcomers and quite a few people in the MP network, but only two Shadbush members besides me were able to come. One of the Shadbush members, Casey, was supposed to cofacilitate the meeting along with someone from the Mountain Watershed Association, who ended up being unable to attend, leaving him to facilitate a very difficult meeting alone. The meeting was somewhat disorganized and hard to control, as people broke into private conversations at times and jumped around in their discussion. One person who came to the meeting after attending the Frack Forum was critical of how the meeting was going, saying he wanted to talk about outreach and that we needed "a serious, strategic political meeting on how we are going to reach people." Another person also expressed impatience that the meeting was skipping from topic to topic, saying "we need to make concrete plans." In the end, however, a fair amount was accomplished because the organizers had previously made some plans and participants volunteered for various tasks such as revising a draft letter from coalition members to the county council, revising a draft pamphlet, planning a press conference for the day of the next county council meeting, organizing people to sign up to testify at that meeting, contacting organizations and other potential allies, and researching council districts (fieldnotes, August 3, 2013). At the next meeting, we selected the name Protect Our Parks (fieldnotes, August 17, 2013). Meetings to plan actions and campaigns can be disorganized, but they also tend to be lively, and tasks involved in organizing actions and campaigns help to focus these meetings.

The campaign was off and running as an important strategic response of the local anti-fracking movement. As Chapter 3 described, POP engaged in a series of tactics and mobilized large numbers of people during the three years of its active existence. POP brought together activists from different organizations with somewhat different tactical experiences and ideologies, but a core group of about ten people became the backbone of the coalition. While Shadbush had initiated the campaign, and continued to participate in various activities, its members (with the exception of me) did not continue to attend most POP meetings. The core members of POP, who met regularly for nearly three years, typically twice a month, were a very congenial group, which included highly active individuals who continued to spearhead new strategies with the backing of the coalition. Although a much larger network of people participated in POP activities such as testimony at county council meetings, collection of names on petitions, and demonstrations, it was the core group that discussed and formulated strategies at POP meetings. Aaron Booz prepared an agenda that was circulated by email in advance of the meetings and he skillfully facilitated most of the meetings; unlike some groups that insisted on rotating facilitators, POP was extremely grateful that he was willing to take on the role. Various strategies and tactics were discussed carefully and talented and committed individuals in the group took responsibility for getting them off the ground. This lasted until core members felt that they had done as much as they could do; although unable to save Deer Lakes Park from fracking, POP may have succeeded in discouraging any move to sell the drilling rights to other county parks. In addition, POP helped to develop the movement community, introducing participants to new tactics and targets, and its strategies were later adopted by other groups, such as the new Allegheny County Chapter of the Izaak Walton League (see Chapter 6).

THE PITFALLS OF STRATEGIC PLANNING

As the above descriptions of efforts at strategic planning show, it is no easy matter for grassroots groups to devise strategies. External threats and opportunities, such as the possibility of fracking in county parks and the seeming unpopularity of the county executive, provide impetuses for movement responses aimed at specific targets. But focused campaigns in response to the actions of opponents are not automatic. Moreover, external opportunities and threats may shift quickly, requiring movement actors to adjust. For example, DEP director Michael Krancer looked like an excellent target, but before a campaign got underway, he announced his resignation as of April 15, 2013

(Hopey 2013b). To build a movement and have an impact, activists need to keep devising strategies and tactics in different political environments and they need to create effective forums for doing so. It turns out, however, to be very difficult to have meaningful discussions of strategy in all kinds of different forums. Many interesting ideas are generated, but few are followed through on. Special strategic planning meetings and workshops are held to allow for extended discussion and to train activists in how to make strategic choices, but their impacts are often uncertain.

At workshops, participants often gain familiarity with tools such as the Midwest Academy strategy chart, but they frequently want more connection to specific movement campaigns and concerns. In reflecting on the Shadbush direct action workshop, Jean Marie commented that "it just could have been more specific to actual actions people wanted to do" (interview, April 25, 2015). At the strategy session offered by the Food and Water Watch staffer, similarly, participants wanted to be able to apply the information to a campaign around fracking as part of the workshop. Educational activities that do not directly lead to anything are seen as less valuable than those done for a specific purpose. Shadbush's effort to conclude the pipeline counter-summit with a strategy session was an attempt to connect the educational event to strategic action, but it underestimated the obstacles to devising strategy in that forum.

Just as participants want to see connection between strategic discussions and current movement campaigns, they need guidance in how to discuss strategies. Workshops such as the one offered by Shadbush on strategy are useful in this regard, but if they are not immediately connected to a particular effort to come up with a campaign, it is hard to judge their impact. A session in which a knowledgeable organizer guided the group on planning an actual campaign would no doubt be extremely helpful. The strategy sessions that led to the creation of POP were successful because they were connected to a real campaign, though there might have been more direction from organizers in the process. As described above, the first open meeting to organize the public parks campaign was somewhat chaotic. Casey's experience facilitating an unruly meeting had lasting implications in that it added to Shadbush's later perception that it was difficult to participate in POP in part because their style of meeting was so different. While the meeting ended up being productive, it might have gone more smoothly with additional planning and more guidance on aspects of the campaign that needed to be discussed. Because activists often have very different ideas about what it means to discuss strategy and plan a campaign, organizers need to provide a clear format for the discussion of strategy at open meetings. This is rarely done, it seems, because grassroots activists value participatory democracy and fear being overly directive.

Follow-Up

When strategic ideas related to movement goals were generated, or when actions were carried out, the lack of follow-up remained a problem for all of the groups I observed. There were different types of follow-up problems: Very often, ideas were thrown out at meetings, but no further action was taken. On some occasions, a decision was made to act on an idea but it was never carried out. For example, Shadbush decided to organize a meeting with their county councilor as part of the POP campaign, but the member who volunteered to lead this plan simply never found the time to do so (fieldnotes, December 17, 2013). Sometimes actions were carried out, but no additional actions were taken to follow up on the original ones, making them in some instances less effective than they might have been. In another example, AG members recalled meetings with public officials as part of the Sierra Club's "cool cities" campaign, but no follow-up discussions (fieldnotes, December 13, 2014). In other cases, actions were intended as one-off events and the follow-up needed was an evaluation of their impact; this type of follow-up failure was most likely in groups that did not have a strong culture of evaluation. In other cases, the lack of follow-up was typically not due to lack of awareness of the problem, but more often to a lack of capacity. Grassroots groups often do not have the resources and structures needed to implement strategic ideas. In particular, they often do not have enough organizers with the time and energy needed to transform many strategic ideas into actions and campaigns.

Collective action typically requires motivated individuals who take the lead in organizing particular activities. Organizational structures, such as an outreach committee or a media group, can be useful, but often must be formed at the time of a campaign, and they need individuals willing to take responsibility for organizing them. Grassroots environmental groups can be very active when they have a supply of individuals who are willing and able to spearhead actions and campaigns. Sometimes, however, enthusiastic individuals take on more than they can realistically handle, or their life circumstances change so that they can no longer manage responsibilities. This happened in Transition when one of the young organizers who was planning to help reorganize the group found himself lacking the time needed for this task. Groups too may over commit their resources, and strategic planning sessions, ironically, sometimes encourage them to do so. For example, at its third retreat, some members of Shadbush were enthusiastic about intersectional organizing, but the group did not actually have the capacity for this. As Patrick reflected, "one of the drawbacks of doing those retreats is sometimes we're not really honest with ourselves" (interview, April 12, 2015).

Some groups have internal cultures that encourage follow-up, though they may still lack capacity. This was true for Shadbush, which had a culture of individual responsibility and extensive participation that encouraged follow-up in various ways. The group always took notes at its meetings, and individuals often kept track of ongoing discussions when they created an agenda for the next meeting (though sometimes this was not completed in advance). Individuals often asked other participants at meetings who would complete a particular task and usually someone would volunteer, though sometimes no one did. When actions were planned, working committees were typically formed and, following an action, the group almost always scheduled a discussion to evaluate the action and think about changes needed for subsequent actions. These were extended discussions that did influence the group's subsequent decisions. The retreats were occasions for setting goals, which were recorded, and continually consulted – in part because particular individuals regularly reminded others about the goals. And Shadbush did follow through on many of its goals. At the first retreat, for example, the group decided to try to work with community groups and to spread direct-action tactics. Both were accomplished in its 2012 action with MOB at the office of state legislator Brian Ellis, and MOB and Shadbush developed a strong relationship as a result of carrying out this tactic. MOB members later participated in the Shadbush action camp and in the memorable protest with the giant pig at the nearby drilling site (see Chapter 3). As Meyer and Kimeldorf (2015) show, participation in tactics can be a transformational experience that helps to build a movement community. Shadbush played a remarkable role in initiating direct-action tactics but could not always follow through on its strategic plans because, despite having some very committed members, the group was a small collective with limited capacity.

The Allegheny Group of the Sierra Club did not have the same type of internal culture as Shadbush, but the group did try to encourage individual responsibility and to devise strategic plans and follow up on them. Despite their best efforts, however, group leaders were disappointed that the strategic planning sessions did not result in much change in the group. Again, one of the big problems was capacity; both issue committees and action teams lacked enough members to carry out many actions aimed at achieving group goals. The AG clearly needed to build capacity, but an ongoing dilemma is that it takes some capacity to build capacity. While the group participated in many coalition efforts, it was having trouble creating its own campaigns and the strategic planning sessions were not solving the problem. One deficit was a lack of ongoing strategic discussion. In contrast to the extensive discussions in Shadbush analyzing all of the group's activities, such discussions were muted in the AG. For example, when the chair asked for comments on the recent strategy session at an ExCom meeting, no one said anything until I suggested having some ongoing discussion at ExCom meetings and others then agreed that this follow-up would be useful (fieldnotes, February 11, 2013).

CONCLUSION

Strategies and tactics are critical to the growth and effectiveness of grassroots environmental movements. In devising strategies, grassroots groups are typically committed to a participatory democratic process, and they value diversity.

Diverse activists are critical to the spread of the movement and they generate many strategic ideas, but groups do not always have the ability to implement these ideas. While participatory democracy encourages innovation and commitment (Polletta 2002), some of the rituals of participation (e.g., icebreakers, exercises, and small groups) eat up time at meetings that might otherwise be used to figure out which ideas to adopt and how to implement them. Such rituals are more useful when they are related to the development of strategy and relationships in the group. Moreover, diverse participants often have varying ideas about how to go about discussing strategy, and workshops on how to create strategy are not always closely connected to actual strategic decision-making processes. While leadership need not be hierarchical, organizers could provide more guidance to strategic deliberations. Leaders or organizers are also needed to follow through on strategic plans, but groups may lack the capacity to organize actions and mount campaigns even when they have leadership. It takes a lot of meetings to plan strategy sessions and follow up on them, and volunteers in grassroots groups that lack staff are often overworked.

To develop their capacity, groups need to create structures, such as recruitment and media committees, that assist in implementing strategic plans. Some working groups can be formed as part of a campaign, but a few ongoing standing committees would likely be useful to "combat readiness" (Gamson 1990). Perhaps most importantly, groups need individuals who will take the lead in organizing actions and campaigns. Ideally, there should be a diverse leadership team and forums for regular interaction and deliberation (Ganz 2000). As we have seen, however, it is not so easy to create and maintain such forums and teams. Teams such as the Marcellus Protest Coordinating Committee are often unsure of their authority. In the Sierra Club ExCom, an elected body, authority was clearer, but deliberation was not as extensive as in Shadbush, a group with a collective culture. The AG had some standing committees such as a fundraising committee, but it needed a strong recruitment committee. While all of the groups I observed shared some of the same problems in devising strategies, these problems also differed across organizational structures.

This chapter, together with the previous one, demonstrates how difficult it is for grassroots organizations to devise and carry out strategic actions and campaigns. Nevertheless, these groups have pulled off many strategies and tactics. And, despite the decline of some organizations, the social movement community endures. Transition Pittsburgh, one of the shortest-lived groups in my sample, organized sporadically for only a few years, but spread its ideas to a network of supporters and affiliates, such as Sustainable Monroeville, that remain active. The Allegheny Group of the Sierra Club, at the other extreme, meets regularly and remains active some fifty years after its founding. Marcellus Protest organized local residents for about three years to carry out numerous activities and to kick start the regional movement against fracking. After the group stopped meeting, activists remained connected through on and offline networks, and many continue to be involved in the larger movement. Shadbush was active for nearly four years and carried out an amazing range of actions despite its small size. Protect Our Parks, which was an outgrowth of Marcellus Protest and Shadbush, raised consciousness about fracking and the need to protect county parks for three years. Although grassroots groups face many challenges, and may not survive long as organizations, they energize the environmental movement community by educating and involving local activists.

9

Conclusion

Grassroots activism is essential to the contemporary environmental movement. The danger to our planet from climate change is urgent and overwhelming, and related threats to water, air, land, endangered species, and human health are immense. In a situation of "all hands on deck," local activists as well as state, national, and international organizations have much to contribute. Case studies of local movements reveal that grassroots environmentalists have achieved a great deal over many decades by raising public consciousness, changing popular discourse, testifying at hearings, lobbying government, and engaging in direct action (e.g., Bell 2013, 2016; Bevington 2009; Bullard 1990; Ferguson 2015; Longhurst 2010; McGurty 2007; Pellow 2014; Walsh et al. 1997). To be sure, there are limitations on the ability of movements to have an impact, and in some places, movements never organize at all to combat environmental threats (McAdam and Boudet 2012). My study of the Pittsburgh movement uses fieldwork conducted over a number of years, following several organizations to the end of their life spans, to provide a close-up look at grassroots environmental groups. This inside description reveals both the excitement and the messiness of collective action, which helps explain the limitations as well as the accomplishments of the movement. In this concluding chapter, I outline the implications of my research for the study for social movements and for movement activists and organizations seeking to create change through grassroots environmentalism.

CONTRIBUTIONS TO SOCIAL MOVEMENT THEORY

Social movement scholars have created a rich body of empirical and theoretical work on social movements and collective action. Over thirty years ago, McAdam et al. (1988) published an influential review of the social movement field in which they found that much scholarship focused on the emergence of social movements and that more research was needed on the dynamics of movements over time. They argued that a lot of the action of social movements takes place at the organizational or "meso" level of analysis, including strategic decision-making within movement organizations. Noting that "we haven't devoted a lot of attention to the *ongoing accomplishment of collective action*," McAdam et al. (1988:729) concluded that "what is needed is more systematic, qualitative fieldwork into the dynamics of collective action at the intermediate meso level." This struck me as exactly the type of research I wanted to do, and my study of grassroots environmentalism in Pittsburgh finally gave me the opportunity to examine movement dynamics over time through extensive fieldwork. In the meantime, some very good studies have taken us inside movement organizations and looked at some of the ways in which movements develop and accomplish collective action (e.g., Blee 2012; Ganz 2009; Han 2014; Lichterman 1996; Polletta 2002). Chapter 2 summarizes some of the work relevant to this study that guided my thinking about how movements mobilize and act. My study contributes to this literature by looking closely at what happens within movement organizations and coalition efforts, elaborating on why grassroots activists experience both successes and difficulties as they create organizations, attract participants, and devise strategies and tactics.

Using fieldwork methods to study multiple organizations and activities in the movement community over time, I was able to examine many aspects of a grassroots movement: individual motivations and participation, organizational structures and cultures, relationships in movement communities, and strategies and tactics, including framing. I show that a variety of people who engage in various forms of activism and join different organizations become grassroots environmentalists. As described in Chapter 3, the groups in my study attracted green entrepreneurs, liberals, radicals, seasoned activists, newcomers to environmentalism, rural and urban residents of threatened areas, and other concerned citizens. They were motivated by threats to their communities, concerns about health, worries about air, water, and land pollution, endangered species, love of nature and outdoor activities, and a desire to save the planet. They engaged in a wide range of strategies and tactics such as educational work, sustainable living projects, electoral politics, demonstrations, direct actions, lobbying, testifying, and petitioning. Although environmental movement organizations need and want to diversify further, particularly in terms of race and class, my study reveals that the movement attracts a variety of grassroots activists and groups. This diversity makes for a vibrant movement even as it sometimes creates challenges within organizations and alliances.

Close-up study adds to our understanding of the complicated process of mobilization. Chapter 4 showed that numerous environmental issues have roused local residents to donate their time and resources to the movement over the years. During my study, threats such as the possibility of fracking in one's neighborhood provided strong incentives for participation in movement demonstrations and other tactics. But even highly motivated people needed guidance on how to organize, and experienced activists were important in assisting newcomers and creating new organizations such as Marcellus Protest (MP). An ongoing local environmental movement, which included long-term activists and organizations such as the Sierra Club and the Group against Smog and Pollution as well as new activities such as protests against mountaintop removal, was critical to movement growth. Framing was also important in reaching a variety of constituents, and movement actors worked hard to present issues in ways that captured the experiences and grievances of people by collecting their stories, organizing educational events, and visiting affected communities. The anti-fracking movement drew on existing community and environmental rights frames and also generated new frames to counter industry discourse and explain movement goals. Yet my observations show how uncertain the process was, even when naming organizations, which is part of framing work. Activists cannot control framing in a diverse movement, and they cannot predict all of the impacts of their framing choices, no matter how carefully they may try to do so.

Collective action is critical to movement building and influence; movement organizers have to generate ongoing collective actions and campaigns to achieve goals and to grow and maintain a movement. As described in Chapter 5, grassroots environmentalists mobilized participants using a variety of mobilizing tactics and campaigns. Supporters became committed activists when they found ways to become involved and when they developed relationships in movement organizations and communities that promoted collective identity. Collective action campaigns and tactics generate solidarity, maintain involvement, and bring in new participants, but it is difficult for organizers to continually devise new actions over time. Quite a lot of time and energy goes into movement actions, and activists often get burned out and need to turn their attention to family or jobs after a time. It is also time-consuming to recruit new activists who might provide fresh energy to the group.

Watching activities and organizations in the Pittsburgh movement over time, I was able to see how different groups fared as well as how the local movement community evolved. To keep up the momentum created by threats and motivated activists, organizational infrastructure and leadership teams are needed. My study suggests that there is no one model of grassroots organizing that maximizes participation and the ability to generate ongoing collective actions and campaigns. All movement organizations need to recruit and involve participants on an ongoing basis and regularly devise strategies and campaigns. They all need to hold some type of meetings and strategic planning sessions to attract participants and devise strategies. They need committees and working groups that allow members to get involved in maintaining the group and engaging in actions. As shown in Chapter 6, all five groups in my study energized individuals to contribute time and resources to the movement for a time and created relationships that supported organizations and campaigns. When particular organizations and campaigns ended, individuals often remained connected to other individuals, groups, and activities; relationships in the movement community and repertoires of collective action maintained the movement. While all of the groups struggled to develop new strategies and tactics and to keep members involved, different types of structures helped with various problems.

Leadership teams enhance strategic capacity, and supportive structures are needed to create and maintain effective leadership teams (Ganz 2000). My research shows, however, that it is more difficult in practice than in theory to create such a team. In my sample, the Allegheny Group (AG) of the Sierra Club had the clearest advantage in this regard owing to a structure provided by the national Sierra Club that featured elected executive committee (ExCom) members and other positions such as issue chairs. This structure helps to account for the longevity of the group since its founding in 1970, and it is no accident that the AG is the one group in my sample that continues to meet regularly. An elected chair and board with terms of several years provide much organizational stability. Yet the structure did not solve all of the group's leadership problems. The AG sometimes had trouble attracting enough active members to the ExCom and occasionally had to replace members who failed to attend meetings and were unable to commit time to the group. In this regard, the MP Coordinating Committee was similar in that it had some committed participants and functioned reasonably well for a time, but eventually struggled to find enough people to serve on the committee. Unlike the AG's elected ExCom, Marcellus Protest's CC had no chair and would not have been able to ask members to serve multiyear terms, though it was useful to request six-month commitments. MP also lacked internal agreement about what powers the CC should have and whether or not there should be elected leaders or paid staff (though the group lacked the resources to hire staff). Transition failed to create a leadership team despite an effort to do so because there was no core group or set of entrepreneurs in place to create the needed structure; the organizing model provided by the international Transition organization was never intended to create a stable organization. The Shadbush collective functioned well for a time as an informal leadership team because it was small enough and members were committed enough to make decisions and carry out projects together. This did not always work perfectly, as some unrealistic decisions were made, and some projects could not be completed. Shadbush experienced ebbs and flows of activity and ultimately disbanded when too many valuable members departed. Protect Our Parks (POP) was similar to Shadbush in that it maintained a core of committed people who worked well together. It had the added advantage of a regular facilitator who prepared an agenda, with emailed input from the group, in advance of meetings. But POP was created around the specific issue of fracking in the county parks, and the group stopped meeting when participants felt they had done as much as they could do at the time.

The ability to create a leadership team is closely connected to the ability to attract committed members to a movement organization. Recruitment requires the development of relationships, the creation of roles and structures that allow participants to learn skills and feel like their time is well spent, and the prospect of collective action. Actions and campaigns are critical; participants develop collective identity in the process of planning for and engaging in collective action. When external threats and events stir up emotions, an ad hoc group can organize protests such as the November 3, 2010 demonstration that gave birth to Marcellus Protest, but it is harder to continue to generate actions. The problem many groups have is that, while actions and campaigns mobilize people, a sufficient number of activists are needed to create actions and campaigns in the first place. Organizations need mobilizing structures, which might take the form of committees and meetings, but it is hard to maintain these structures without exciting actions. In addition to the lack of structures such as recruitment committees, existing volunteers in all of the groups I studied were stretched too thin to do the work needed to recruit people. Because it takes a great deal of effort to bring in new volunteers, staff may be better able to do this than volunteers. In the absence of staff, organizations need to deliberately develop the commitments of volunteers (Han 2014) and to create structures that allow volunteers to take on meaningful roles rather than providing only opportunities for "plug-in" activism (Lichterman 2005). At the same time, groups need to attract enough volunteers and to create a division of labor such that a few individuals are not doing all the work – a recipe for burnout.

Most of the groups in my sample struggled with how to create an effective structure or to make the existing structure work. Again, the Sierra Club had the advantage of an established structure, but still had difficulty maintaining active issue committees, particularly as older members became less active and the group failed for a time to recruit enough younger members. My observations revealed that the group structure and core members, who developed strong relationships with one another, kept the Allegheny Group going. Although it was difficult for the AG to create new campaigns, members sometimes introduced new activities, such as the successful pollinator garden project in Wilkinsburg that began in summer 2017. The AG frequently participated in coalitions such as POP and the Clean Rivers Campaign as well as regular activities such as electoral campaign work. Staff members continued to involve AG members in their projects, though their activities were typically organized independently rather than in direct coordination with the group. In addition to hiring staff members and funding a field office, connections to the state and national Sierra Club organizations provided many different types of resource to the local group.

Grassroots groups that have to create structures from scratch face greater challenges, particularly when their members have diverse ideologies, backgrounds, and experiences. In the case of Marcellus Protest, there was uncertainty about how formalized the structure should be, how often working groups should meet, and who had the authority to do what. When diverse groups meet, as shown in Chapters 7 and 8, participants often have different ideas about how to discuss strategy and, lacking strong guidance from organizers, discussions are often unfocused. Most grassroots groups try to be as democratic as possible, and some eschew formal leadership, making it harder to control meetings and to follow up on the wide variety of ideas that are often presented. Diversity is an asset insofar as it provides groups with new resources and perspectives, but the advantages of diversity can easily be squandered when there is a lack of structure for follow-up. As a more ideologically cohesive group, Shadbush had an easier time agreeing on structure and formulating strategies. While "leadership" was rarely if ever discussed, the collective did operate as a team, and individuals often stepped up to perform some of the functions of leadership, such as summing up decisions and making sure tasks were distributed. Members were incredibly committed and accomplished a great deal, but it was difficult to sustain activity with such a small group, which remained homogeneous and was unable to expand its membership and capacity.

The large-scale political and cultural contexts of social movements are critical to their ability to mobilize and effect change (Tarrow 2011). My research demonstrates that the meso or organizational level is also essential to our understanding of movement mobilization and strategy. Activists express their views, learn and use skills, and develop relationships in social movement meetings and collective actions. Chapter 7 described how organizational cultures and structures influence meetings, which in turn affect the ability of the movement to maintain itself and engage in collective actions and campaigns. Social movement meetings are related to collective actions and campaigns in that they are most likely to be successful when focused on planning appealing activities. While meetings are not always interesting, they are needed to plan actions, and they allow participants to form commitments to organizations. Chapter 8 described some of the different types of meetings and forums that are the sites of strategic planning. A close look at what goes on in meetings and workshops shows just how difficult it is to devise strategies while trying to maintain maximum participation. Input from a diverse membership can be a source of strategic innovation and new ideas, but it can also create disarray and disappointment. This occurs when meetings are not well facilitated, when misunderstandings or conflicts arise, and when there is no follow-up on the ideas generated. Grassroots groups typically strive to be democratic and participatory, but these goals may lead to a lack of sufficient guidance by organizers or leaders as groups try to create actions and campaigns. Leadership teams are needed, but my observations showed little discussion of leadership and how to create such teams.

While internal organizational dynamics are critical, my study also reveals the importance of relationships in the larger movement community in maintaining grassroots environmentalism. Alliances are an important source of ongoing movement activity, and networks among individuals and organizations can lead to the creation of alliances such as POP. Even when the Sierra Club initiated few exciting activities of its own, the AG participated in alliances, often bringing out a number of members for rallies and tactics such as testimony to the county council. Although activists identify strongly as Sierra Club members, they have been quite willing to work with other organizations and in coalitions such as Pittsburgh United, which brings together labor, community, faith, and environmental organizations. Marcellus Protest ceased to meet as an organization in 2013, but activists maintained contact through Google Groups and social media as well as personal networks, and individuals continued to participate in various movement events. In 2019, core members of POP and MP got together to talk about taking actions together. POP tactics were one of the inspirations for the formation of a new chapter of the Isaac Walton League, headed by former MP and POP activist Mike Stout. Transition members can also be seen at movement gatherings, such as Maren Cooke's Sustainability Salon, and at meetings of organizations such as Pittsburgh 350.org. Relationships between former Shadbush members and organizations such as Marcellus Outreach Butler and the Center for Coalfield Justice have endured. Once a grassroots movement mobilizes, it continues to have an impact even if individual organizations decline (Wood et al. 2017).

LESSONS FOR MOVEMENT ORGANIZERS

Grassroots movement activists and organizations can accomplish a great deal. The groups that I studied were by no means failures, though each had its limitations. One lesson for activists is the need to carefully consider different choices regarding organizational structures and how they affect longevity and strategic capacity. While few grassroots groups seem to spend much time thinking about how to create leadership teams, this is a critical component of structure. Leadership does not have to be hierarchical, but practices such as relationship building, skill development, and strategic planning need to be accomplished (Ganz and McKenna 2019). When a team of people accountable to the organization takes charge of accomplishing these tasks, the group is likely to be longer lived and more effective. But another quite different lesson is that movements have impacts beyond the boundaries and lifetimes of individual movement organizations and campaigns. While campaigns end, and organizations do not always survive for more than a few years, movements typically continue indefinitely. The movement-building work of organizers is important, and outcomes such as new public consciousness and discourse, stronger and more diverse networks in the movement community, and new tactics in the local repertoire of collective action should be recognized as successes for a grassroots movement.

The groups in my study illustrate the strengths and weaknesses of different types of movement organizations. The Sierra Club is a stable organization with a conventional structure that does important work such as lobbying and campaign work, but during the period of my observations had trouble expanding its active core and developing energetic new campaigns. Transition Pittsburgh generated some innovative ideas and activated some members, but had trouble following through on its ideas and could not create a stable organization. However, the ideas remain important, and individuals spread them in other venues. Marcellus Protest created consciousness about fracking and helped build the movement with its actions, website, and social media. Although the group was unable to create a lasting organizational structure, it did create an ongoing network. Shadbush also played a key role in raising consciousness about the dangers of fracking and other environmental problems, demonstrating the potential of an ideologically cohesive collective as well as its problems in expanding and diversifying. Protect Our Parks was a capable group that generated a lively campaign and may have prevented leasing of more county parks, even though it failed to stop fracking in Deer Lakes Park. The group might have changed its focus to continue meeting after it stopped doing so in 2016 – and at least one Marcellus Protest activist raised the possibility of reviving the larger group before POP disbanded. However, my sense was that the core members of POP were somewhat weary after meeting about twice a month for three years, usually on Saturday mornings, and most felt that the group had arrived at a natural end.

Most of the groups in my study failed to develop leadership teams and lacked adequate structure for continually mounting projects and campaigns and recruiting new activists. In the case of the Sierra Club's AG, which had more leadership and structure than the others, I was surprised that staff employed by the national organization or the state chapter tended to work parallel to the local group rather than working directly with it. While the staff no doubt engaged in important work in the community and helped the AG when possible, their jobs did not include direct aid to the group. More integration of staff into the local group – for example, having a staff position on the ExCom – might strengthen the AG leadership team and provide volunteers with more organizing expertise. If mandated to do so by the national or state Sierra Club, staff could provide leadership training and teach community organizing skills to volunteer members of the ExCom as part of their jobs. ExCom members would then be better equipped to recruit more members and to develop their skills and commitments. In other groups, there was little discussion of leadership and how participants might be trained to take leadership. Indeed, there were typically few, if any, formal positions for volunteers to take.

Ongoing actions and campaigns were necessary for integrating new activists, but most groups relied on individual initiative rather than organizational structure to create the activities that were critical to continued mobilization and recruitment. In Marcellus Protest, for example, individuals took it upon themselves to organize small demonstrations, but the Action committee never became a working group that met regularly to think about strategy and organize actions. Without such a structure, it is difficult for groups to continue their activity when individuals can no longer take the lead. This typically happens when individuals burn out from too much movement activity and/or when work or family responsibilities curtail their time. A committee structure creates more of a leadership team, and also a place for new volunteers to become involved and develop their skills. A leadership team helps solve the problem of burnout that occurs when individuals do too much and cannot sustain the high level of activity. A team can also provide an ongoing source of leaders as new members continually join the team. While it can be difficult to maintain committees, particularly in the absence of an action or campaign, grassroots groups are most likely to thrive when they have both campaign-based committees, such as a logistics committee for a demonstration, and standing committees, such as a fundraising or recruitment committee. In the case of Shadbush, the collective organized committees when needed for an action, but might have brought in other members with standing committees that provided ways for more people to become involved.

Meetings such as general assemblies and forums for strategic planning are also places for people to become involved and where excitement about the movement can be generated. Those that are directly connected to group actions are often more valuable than educational forums and training workshops that lack a specific purpose. The MP general assemblies leading up to the November 3, 2010 demonstration provide a good example. Although not all of the meetings were equally engaging and productive, most of them generated enthusiasm and made participants feel that the movement was going to have an impact. The demonstration provided a focus for the organizing by a diverse group of people, and experienced organizers helped guide the meetings. The meeting after November 3 was lively, but also disorderly, and it was clearly going to be another matter to keep the group – and the movement – going. While a bureaucratic structure might, as Piven and Cloward (1977) warned, stifle a movement, grassroots groups need to figure out ways to increase strategic capacity by creating workable structures that promote democratic participation and accountable leadership.

Diversity in a group can make organizing difficult as people often have different ideas about how to go about organizing depending on their experiences and backgrounds. The lesson from this is not that diversity is undesirable; diverse constituents are important in expanding movements and bringing new strategic ideas and resources to movement organizations (Ganz 2000; Polletta 1999). But groups need to recognize and discuss different ideas about how to organize and different understandings of what it means to plan strategy and think about goals. Diversity can make strategic deliberations difficult owing to differences in style and contrasting perceptions of what strategy means. Meetings are important for bringing activists together, but they don't appeal to everyone. Groups do need to meet regularly, but meetings planned around actions or that result in actions or campaigns are typically the most energetic.

Participants get frustrated when meetings generate good ideas without follow-up, and this often happens in the absence of leadership teams and adequate organizational structures. Strategic planning sessions help groups think creatively about goals and tactics, but they are most useful when they are connected to the planning of real tactics and campaigns. Leaders and organizers need to provide direction to strategic discussions and ensure that follow-up occurs.

Grassroots action is exciting at its best, and environmentalism needs the energy of local activists, who experience environmental problems on the ground. I was amazed at how much the groups in my study accomplished, but I was also struck by how difficult and messy the process of organizing is up close. By looking closely at the sources of those difficulties, I hope to have contributed to both social movement theory and the practical politics of environmentalism.

Fieldwork Appendix

Studying Movement Dynamics

Organizational dynamics and the nitty-gritty of how movements mobilize have always interested me greatly as a scholar of social movements. Participant observation affords the researcher a close-up look at these dynamics, but it is by no means a straightforward method with clear procedures and rules. Participant observation in social movements is difficult, time-consuming, and inefficient, and there is no guarantee of theoretical payoff. But it can also be highly rewarding in terms of insights gained by sharing experiences with activists, talking to people informally, and watching strategies unfold. In the following account, I discuss the challenges and rewards of the fieldwork that I conducted for this study, including many hours of participant observation and supplementary interviews with activists in the groups that I studied.

SELECTING FIELD SITES AND GAINING ACCESS

When I began this project in 2010, I knew that I wanted to study the environmental movement using participant observation. I have long been concerned about environmental issues – probably since Earth Day 1970 raised my awareness as a high school student – and I wanted to contribute to understanding how the movement can be effective in the face of climate change and other urgent problems. I was interested in observing grassroots groups and activities because this is where active participation occurs. I wanted to see how the movement engaged local volunteers and developed strategies and tactics beyond the paper membership and lobbying work often done by national environmental organizations. Because I wanted to do fairly long-term participant observation, I decided to study the movement in Pittsburgh, where I live, as a practical option. (At one time, I had thoughts about a multisite study, but the demands of my methods would have required me to gain funding and build a research team to accomplish this, and I ended up deciding that a single-site, multiorganizational study would yield significant theoretical insights.) Although I was open to what I might learn inductively, I began with some theoretical ideas (discussed in Chapter 2), and I wanted to sample groups with a range of organizational structures and ideological approaches. I also wanted to take advantage of the current activities in the local environmental movement so that I could see strategic planning and collective action as it happened. I was fortunate to begin my study at an exciting time for the movement, just as protests against fracking were picking up steam in Pittsburgh and Pennsylvania.

Initially, I kept my eye out for environmental groups and activities of all types, and I went to as many public events as possible. I had met some local activists and joined some email lists during my previous study of the 2009 G20 protests in Pittsburgh, and an email from the Thomas Merton Center, a local peace and justice movement center, alerted me to a public meeting to plan a demonstration against fracking. An activist I met through the G20 protests forwarded me an email about Transition Pittsburgh's showing of the internationally produced Transition 101 film. I went to these and other environmental events, such as a talk on mountaintop removal, and thought about what I wanted to study and how I wanted to structure my research. I wanted to vary the groups I studied by ideology and structure, and I wanted to study groups working on currently pressing issues. Because the exciting activity was happening around fracking, I quickly became involved in attending Marcellus Protest (MP) meetings to organize a demonstration on November 3, 2010. Although researchers might prefer to get permission to study an organization *before* beginning a study, participant observation requires a process of figuring out what to study and gaining access to groups and their activities. Moreover, emerging movements such as the anti-fracking movement, and the group that became Marcellus Protest, are typically not equipped to approve requests from researchers when they are first organizing.

I decided to begin my study by contributing what I could to the effort to organize the November 3, 2010 protest against fracking. In private conversations, I told some individuals that I was a sociologist interested in studying the local environmental movement, while also noting truthfully that I was in complete sympathy with the movement. I explained that this was an exploratory phase of my research, and no one seemed the least bit concerned about this; rather, they were interested and pleased that I was participating in the movement. I joined the Outreach committee formed to reach out to individuals and organizations in preparation for the demonstration and contributed in small ways such as helping to draft a letter to organizations asking for their endorsements. I took fieldnotes on all of the meetings and events that I attended and looked for an opportunity to formally ask the group's permission to act as a participant observer. I talked to Jessica, who was scheduled to facilitate the first meeting following the November 3 demonstration, and we agreed that the agenda for that meeting was much too full for me to explain my project then. Instead, I sent an email to regular activists in MP explaining my intentions, asking for their feedback, and expressing my hope that we could discuss my research project at a future meeting when there was time to do so. The replies that I received were all supportive of my research intentions.

It was not until the January 8, 2011, MP general assembly that I succeeded in getting on the agenda after communicating with Dana, the meeting facilitator, in advance. At the meeting, I explained that I am a sociologist at the University of Pittsburgh who studies social movements, and that I wanted to do a study of the environmental movement in Pittsburgh, including but not limited to Marcellus Protest. I explained that I am interested in studying how the movement organizes and has an impact, and that I hope to study aspects of the movement that are useful to the group and would be happy to share my findings with them. Some of the people present knew me fairly well, and Mel jokingly asked if I would be collecting information on their sex lives, and I joked back, "yes, detailed information." Gloria quipped, "that will be boring," and we all laughed. Others asked serious questions, such as where I would publish my results, and I replied that it would most likely be in a book and/or sociological journal articles. One woman, whom I did not know, said she would not want her name used, and Dana, whom I'd briefed in advance, jumped in and said I would not use names without permission, and I nodded in agreement. Dana said that if there were any concerns later, they could be brought to the Coordinating Committee (CC) or to me directly. She said, "Be aware that you are being observed." I responded by assuring them that I am a supporter and wanted to continue to be a participant but that I would be learning more about the challenges facing the group as an inside observer and would later ask people for interviews. Mel said, in a supportive way, that I have been participating and contributing to the group for some time. Dana said that everything would be transparent, as I had told her previously, and I agreed. The group agreed that I was welcome to conduct my study (fieldnotes, January 8, 2011).

As Mel's support illustrated, it was clearly important that I got to know a number of people in the group before asking for formal access. However, some of those at the meeting where I asked permission to study the group were relative newcomers who did not know me, and it was also important to explain my project clearly and honestly. Moreover, new people continue to come to a group like Marcellus Protest, so I needed to explain my research to them whenever I had the opportunity. I did this privately when I met new people, and when introductions were made at the start of a meeting, I often explained my role as a researcher. My continued participation was also important in building trust – and I took that trust very seriously as I conducted my research and wrote about the movement. I regularly attended meetings of the new Outreach committee that was formed at the first general assembly held by Marcellus Protest after the November 3 event, and I got to know other participants fairly well. I also attended many other meetings, including a few meetings of the Action committee (which met infrequently), Coordinating Committee meetings, Strategic Messaging committee meetings, and general assemblies.

When MP stopped meeting, and Protect Our Parks (POP) began, I was already well known in the group and had easy access to POP.

My entrée into other groups was also facilitated by my participation. In the case of Shadbush, it was critical that I got to know two members, Jessica and Ben, through my participation on the MP Outreach committee, which met regularly for about a year. After I had been observing MP for about nine months, I realized that Shadbush would be an interesting comparison group for me to study because it operated as a collective and had a more radical ideology. In the summer of 2011, I asked Jessica if I might join Shadbush, and I remember her replying, "You mean as a researcher or as a member?" and I said both. She said she would check with the group, and Ben later sent me an email about coming to a meeting. I had no problem truthfully agreeing with Shadbush's statement of principles, which was a prerequisite to membership. But because Shadbush was selective in who it allowed to join the collective, I'm sure I would never have gotten access to the group if I hadn't gotten to know Ben and Jessica before asking to become a member. But not everyone in the collective knew me, and at the end of the first meeting I attended, I asked Ben if the group was comfortable with my research role and offered to answer any further questions they had. He said he personally was comfortable with it, but he would ask the others. Ben later told me, somewhat apologetically, that a few other members wanted me to answer more questions about my research before allowing me into the group. I did so at the next meeting and was then granted access and added to the member email list.

Transition Pittsburgh was interesting to me as a group with a different type of mission than most environmental groups; rather than engaging in protest and political action, Transition wanted to create models of sustainable living at the community level. Intrigued by this approach, I attended a number of events and got to know a few regular participants before declaring my research intentions. The group was informal – and the membership very fluid – making it difficult to ask for a decision on my participation. However, I spoke about my research intentions at a meeting, and everyone present was agreeable, even enthusiastic, about my study. I participated in discussions at meetings and events and was always warmly welcomed by other members. I kept informing newcomers of my role whenever possible, usually during introductions at the start of events and meetings. Regular members were clearly aware of my research role as they often asked me about my study or mentioned to others that I was doing a study.

I wanted to include the local Sierra Club group in my study as an affiliate of a national organization with an established organizational structure and a more mainstream ideological approach within the environmental movement. I initially had some difficulty figuring out what to observe. I went to some meetings at the Sierra Club office when the Allegheny Group invited Marcellus Protest and others to participate in the St. Patrick's Day parade in 2011. Later that year, I went to the Allegheny Group's holiday party and asked about attending meetings but was told that general membership meetings were no longer being held. Eventually, I talked to a member who said I could come to Executive Committee (ExCom) meetings as long as I was a Sierra Club member, which I was. In April 2012, I went to my first ExCom meeting and explained my project and desire to be a participant observer in the group. Everyone was very friendly and interested, and they readily approved my research plans. They asked some questions, such as where the research might be published, and seemed pleased that I would likely write a book. Because there were regular attenders at the ExCom meetings, I typically did not have to keep explaining my project, but I did so during introductions whenever there were newcomers and also to remind everyone of my research role on occasion.

FIELDWORK

In all of the groups that I studied, I tried to strike a balance between participation and observation. When I attended meetings, I took notes, either by hand or on my laptop computer, which was not only useful to me, but also a reminder to others that I was doing research. As soon as possible after the meetings, I expanded these notes based on memory into fieldnotes. In Shadbush, I was asked to agree to stop taking notes when something sensitive was discussed, such as possible use of illegal tactics, but this was quite rare, and I think there was only one occasion when I was asked to stop taking notes. I was first asked to do minutes for MP general assemblies because others could see that I was taking notes anyway. I was also asked to do minutes for some MP committees, such as Outreach and the Coordinating Committee, and I was occasionally instructed not to put something in the minutes. I sometimes did minutes for the Outreach committee, but two other members, Aaron and Jessica, often volunteered, partly I think to give me a break since they knew that I often did minutes for MP and its CC. I could produce minutes fairly easily by cutting out detail from my own notes, but the process made me realize how "sanitized" minutes tend to be in comparison to fieldnotes! In any case, this became a role for me in MP and later in POP, where I regularly took minutes, and I saw this task as a form of reciprocity – something I could contribute in exchange for access to the groups. I was not asked to do minutes in the Sierra Club, which had its own secretary as part of the ExCom; nor was I typically asked to do minutes in Shadbush or Transition, where young members with laptops regularly produced notes very quickly for their websites or email lists.

Beyond taking notes, I made some comments at meetings, but tried not to dominate discussions. I wasn't particularly worried about influencing the groups unduly because this would have been hard for one person to do in any case, and I considered my own participation and reactions to be part of my data. When I wrote up my fieldnotes, I tried to record my own comments, actions, and feelings as well as those of others. But it would have been hard for me to observe as much, and take good notes, if I was overly active in the groups, and I did not want to be the focus of attention. On the other hand, I often contributed my own opinions, particularly when asked to do so. This helped make me part of the groups I studied, and I also participated in social events, such as potluck meals, and attended protests whenever possible. These shared experiences, and my informal discussions with individuals, were important to my understanding of the movement. Because I was studying multiple organizations, however, I had limited time to contribute to each, beyond attending meetings. My role as a participant observer also varied somewhat in the different groups that I studied. I was most active in Marcellus Protest, in part because the group was quite active during the period of my study, and in part because I fit into the group easily based on my own age and background. This was also true of the Sierra Club, and I joined its Global Warming Action Team and attended some action nights to make phone calls to members, but I did not have time to get involved in other activities such as political campaign work. Shadbush was very welcoming to me, but I was the only older member of the group, and the young members did not seem to expect me to participate in everything they did, not only because of my age but also because of my research role. Older members of Marcellus Outreach Butler did work well with Shadbush, and I certainly could have contributed more to actions such as the "pig" protest if I had the time and inclination. I did not socialize with the Shadbush members after meetings, and they clearly did not expect me to do so. Transition was also very welcoming to me, and I attended as many of their activities as possible. In the case of POP, I went to meetings regularly, and did minutes for the group, but only attended a few actions such as a demonstration at Deer Lakes Park and the final vote on leasing the park at the Allegheny County Council meeting, but these were both memorable and important events. Because other participants saw me at meetings so often, they tended to think I contributed more than I did – and I got a huge amount of credit for doing minutes, which few people wanted to do. While I felt a little guilty at not contributing more, I was pleased that I got to know many activists and gained a good understanding of their concerns and the challenges facing the movement.

In addition to my extensive participant observation, I interviewed a sample of activists from each of the organizations, except POP, where many of the same people were involved, and I had so many informal conversations that I did not think formal interviews were necessary. A research assistant, Corinne Ogrodnik, conducted all but four of the interviews with Transition members, and I conducted the rest of the interviews myself. The interviews were important in allowing key activists to tell their stories and provide insights into movement strategies and outcomes. I interviewed MP activists in 2012, before memories of the early period of activity starting in 2010 faded. Corinne and I interviewed Transition members between November 2012 and May 2014. I interviewed Sierra Club members between December 2012 and June 2013. I saved Shadbush for last, interviewing collective members in the spring of

2015 after the group dissolved in late 2014, allowing members to reflect on their experiences. In all of my interviews, activists were eager to mull over their groups and the movement. It was very useful for me to have experienced meetings and events with my interviewees so that I could ask questions related to what I had observed.

My extensive fieldwork, including participant observation and interviews, presented both challenges and opportunities for theoretical insight. The challenges included a massive amount of data over five organizations and seven years. I kept up with writing fieldnotes, knowing I would lose data if I allowed too much time to elapse between observations and elaborated notes, but it was difficult to keep up with coding the data on the qualitative analysis program I used (MAXQDA). I finally admitted that I couldn't do it all myself and enlisted research assistants to help with the coding. It would have been preferable for me to do all the coding myself, but I went over the work of my assistants, added codes, and wrote memos as often as possible. But analyzing the data was a very slow process, even though I tried to keep thinking about what I was finding and to continually record thoughts in memos. There is a huge temptation with qualitative research to want the data to "speak for themselves," but, in fact, it takes a lot of work to develop theoretical ideas by using and extending the existing literature and by comparing different observations and cases to one another. Although I was well aware of this pitfall, the long-term nature of my fieldwork made me want to see how things unfolded, with the hope of gaining some further insight into mobilization. It was useful to witness strategic choices and the ups and downs of mobilization over time – including the decline of four out of five of the groups I studied – but I still had the job of analyzing what I observed.

My research allowed me to gain understanding of how grassroots movements accomplish what they do and why it is so difficult for them to continue to mobilize and act collectively. By observing groups up close and listening to activists talk, I experienced some of the dynamics of mobilization and collective action. I observed how environmental issues such as fracking motivated all kinds of people to become activists and how much grassroots groups were able to accomplish. In observing recruiting efforts and discussions about the problems of bringing new activists into organizations, I recognized a key dilemma: activities were needed to recruit people, but it was difficult to organize more activities without enough people; it took a great deal of effort to recruit new members, and existing volunteers were stretched thin. I also saw how a lack of structure often made it hard to mount projects and recruit new activists. The difficulties involved in organizing and creating workable structures surfaced in all of the groups that I studied, and I saw how hard it was, for example, to create leadership teams and to hold successful meetings. Regarding strategy, I was able to observe the many obstacles to devising strategies and creating campaigns, which include the difficulties faced by a diverse set of people with different understandings of how to deliberate about goals and strategies.

Watching the movement over time, I saw how relationships among individuals were forming and a movement community was developing, which would endure beyond the decline of particular movement organizations. In short, participant observation allowed me to understand the complexities of grassroots organizing, informing the chief findings of my study.

ETHICAL ISSUES

I applied for and received the "exempt" designation from my university's Institutional Review Board (IRB), which meant that my project was not expected to expose the people I observed and interviewed to any real harm. Interestingly, a few of the activists I encountered were aware of IRBs. For example, Ed, a Sierra Club member and retired university professor who had served on his university's IRB, asked me about IRB approval and ribbed me by expressing the view that my research should never have qualified for exempt status! While I think this was the correct category for my research, I still had to continually think about ethical issues as I conducted the research and wrote about my findings. I was very aware throughout that not everyone I observed knew I was studying the movement, and many who did know this forgot over time, so I tried to keep reminding people that I was a researcher. Fortunately, there were numerous opportunities for doing so, such as the common practice of beginning meetings by having everyone introduce themselves. I typically took this opportunity to say that I am a member of the group and an environmentalist but that I am also a sociologist doing a study of the local movement. When reminded of my research role, individuals sometimes made comments such as "I forgot you are doing a study," but others made clear that they remembered. For example, Barb once explained my role to others at a Sierra Club meeting without my prompting her to do so. On another occasion, after I introduced myself as a sociologist doing a study, Tom said jokingly, "So she's studying you!" Ed looked at me as I was leaving a Sierra Club action group meeting and said in front of the whole group, "You got some good data tonight." While this was a little embarrassing, it was a good sign that some participants understood that I was a researcher and not just an activist.

At the same time, I was continually drawn into more participation in some of the groups. While I intended to attend MP Coordinating Committee meetings largely as an observer, I was persuaded to vote when there were not enough people present for a quorum without me. If no one else was present from the Outreach committee, I was asked to report on the committee's activities. While most of my contributions were minor ones, I was sometimes invited to participate in more extensive ways. For example, I was asked to run for the Sierra Club ExCom, and in declining I explained that I did not want to do so while conducting my study. In Shadbush, my age set me apart from the others to some extent, but I was sometimes clearly designated as a member of the collective. At the action camp held by Shadbush, Patrick gave me a bright green bandana to put around my arm; the collective announced that its members were identified by the bandanas and to feel free to ask us questions. This made me feel a bit like a fraud, since I had done none of the work to plan the action camp and was only helping to register people as they arrived, but it also made me feel identified with the group. Nevertheless, the collective was well aware of my limited role, and they did not expect a lot of action from me.

When I asked people for interviews, they were of course reminded that I was studying the movement. I was always completely honest about my sympathies, but I know that everyone did not understand what it meant for me to be doing research. I explained this as best I could in private conversations with individuals and at the few meetings where I discussed my research intentions. I also tried to be honest when people asked me about myself. This was sometimes difficult, as when Mark suddenly asked me in our interview if I believed in God. I had to decide in a split second whether to be evasive or honest, and I went with honesty, replying simply "no." Mark looked at me with annoyance and later made reference to my lack of belief in a comment, but I suspect he also respected my honesty. People got to know me and trusted me, and I felt that the burden was on me to write about my observations in a constructive way. I didn't want to avoid writing about conflict or problems in the movement, but I tried to do so in a way that helped explain these difficulties and that would not be upsetting to participants.

Sociologists often assume that people should be anonymous in their studies and that organizations should not be named. I always think about studies of the civil rights movement and how they would look if participants and organizations went unnamed. I did not use the names of individuals without their permission, with the exception of a few public figures and persons named in publications or internet profiles. I asked my interviewees whether or not they wanted their names used, and some were quite firm in their assertions that their names be used. I also assured people that I would be cautious in how I used their names and would not do so in a way that might cause embarrassment. This produced some strong-willed responses, such as the comment from one person that "I believe in freedom of speech and so you could use my name for good, bad, or indifferent" and the reply from another that "I may say something embarrassing, in which case it's my problem." Others were uncertain about the use of their names, and some wanted only their first names used. (In some cases, I used first names only for stylistic reasons or to avoid a mix of full and first names when talking about a group or event.) When I was writing the book, I checked back with some people who were previously uncertain, and most felt strongly that with the passage of time, they wanted their names used. They were proud of their work in the movement and were happy to be recognized. I also checked in with a number of people who I had observed frequently at meetings, and talked to informally, but had not formally interviewed to give them a chance to say whether or not they wanted their names used; almost all did.

With regard to organizational names, the groups I studied were all well-known entities in the local environmental movement with unique characteristics that readily identified them. I do not think it would have been possible to disguise their identities and none of them asked me to do so. In the case of Shadbush, a small collective, I originally thought they might want me to give the group a pseudonym – though anyone searching the internet for an environmental justice group in Pittsburgh would quickly locate Shadbush. I discussed use of the Shadbush name with members of the collective on two different occasions; during the first discussion, members laughed about having the opportunity to use previously rejected names such as the "Hellbenders." During the second discussion, we came to a clear agreement that I would use the real name of the group. Patrick commented that he thought it would be "silly not to use the name," and Jessica said, "We can take constructive criticism." Certainly, I felt a strong responsibility to write about all of the groups in my study in a constructive manner, and use of their real names only enhanced my commitment to accuracy and fairness in discussing their experiences. Beyond these ethical issues, I think researchers in the field benefit from knowing the identities of organizations. It allows for the possibility of follow-up research on the organization and its impact in a local movement community. In the cases of the affiliates of national and international groups such as the Sierra Club and Transition, comparisons to studies by other researchers are made possible when organizations are named. Groups such as the Sierra Club have long histories, and an understanding of the larger organizational context adds to our understanding of local dynamics. Moreover, much has been written about such groups by other researchers, journalists, and participants; they can take constructive criticism.

I regret that I never made the time, or devised a forum, for discussing my findings with activist groups while I was doing the study, even though I initially expressed my willingness to do so. I also did not share parts of the manuscript with them as I was writing the book. I did have some discussions with a few individuals, but no collective ones with groups where I might have received feedback from them and perhaps provided advice to them. One reason for this failing was that I didn't feel prepared to talk about my "findings" until I had time to code and analyze my data. My research was a slow process, and I felt embarrassed on a number of occasions when activists commented on how long it was taking me. Before a Shadbush meeting in January 2013, for example, Patrick asked me if I had started writing about the anti-fracking movement yet, and Jessica said, "We're all dying to hear what you have to say." But the tentative ideas I was forming did not seem important enough to present to activists, and I wasn't sure I had anything to teach them. The agendas of most meetings were packed, and I didn't want to take up time even though I knew activists were interested in what I was learning. By the time I started writing the book, most of the groups were no longer meeting, and I hesitated to ask individuals to read the manuscript. I wish I had found ways to discuss my work in progress with movement participants, though I suspect this would have benefitted me more than them. I hope that activists as well as academics find the final product helpful.

References

Allegheny Group of the Sierra Club. 2010. "Endangered Species Action Team Reaches Out to Youth." *Allegheny Sierran: Newsletter of the Allegheny Group of the Sierra Club* 29(1):3.

Almeida, Paul D. 2019. "The Role of Threat in Collective Action." Pp. 43–62 in *The Wiley Blackwell Companion to Social Movements*, 2nd ed., edited by David A. Snow, Sarah A. Soule, Hanspeter Kriesi, and Holly J. McCammon. Oxford: John Wiley & Sons.

Amenta, Edwin, Drew Halfmann, and Michael P. Young. 1999. "The Strategies and Contexts of Social Protest: Political Mediation and the Impact of the Townsend Movement in California." *Mobilization* 4:1–23.

Andrews, Kenneth T. 2004. *Freedom Is a Constant Struggle*. Chicago: The University of Chicago Press.

Andrews, Kenneth T., Marshall Ganz, Matthew Baggetta, Hahrie Han, and Chaeyoon Lim. 2005. "National Purpose, Local Action: Organizational Effectiveness of Sierra Club Groups and Chapters." Report to National Sierra Club.

2010. "Leadership, Membership, and Voice: Civic Associations That Work." *American Journal of Sociology* 115(4):1191–242.

Aunio, Anna-Liisa. 2012. "Mobilizing the Generation Gap: Transnational Coalitions and Insider/Outsider Strategy in the Climate Action Network." Pp. 93–119 in *Strategies for Social Change*, edited by Gregory M. Maney, et al. Minneapolis: University of Minnesota Press.

Aunio, Anna-Liisa, and Suzanne Staggenborg. 2011. "Transnational Linkages and Movement Communities." *Sociology Compass* 5(5):364–75.

Baggetta, Matthew, Chaeyoon Lim, Marshall Louis Ganz, Hahrie Han, and Kenneth Andrews. 2012. "Learning Civic Leadership: Leader Skill Development in the Sierra Club." Pp. 110–138 in *Interest Group Politics*, 8th ed., edited by Allan J. Cigler and Burdett A. Loomis. Los Angeles: Sage.

Baggetta, Matthew, Hahrie Han, and Kenneth T. Andrews. 2013. "Leading Associations: How Individual Characteristics and Team Dynamics Generate Committed Leaders." *American Sociological Review* 78(4):544–73.

Banaszak, Lee Ann. 1996. *Why Movements Succeed or Fail: Opportunity, Culture, and the Struggle for Woman Suffrage*. Princeton, NJ: Princeton University Press.
Beamish, Thomas D., and Amy J. Luebbers. 2009. "Alliance Building across Social Movements: Bridging Difference in a Peace and Justice Coalition." *Social Problems* 56(4):647–76.
Bell, Shannon Elizabeth. 2013. *Our Roots Run Deep as Ironweed: Appalachian Women and the Fight for Environmental Justice*. Urbana, Chicago, and Springfield: University of Illinois Press.
2016. *Fighting King Coal: The Challenges to Micromobilization in Central Appalachia*. Cambridge, MA: MIT Press.
Benford, Robert D. 1993. "Frame Disputes within the Nuclear Disarmament Movement." *Social Forces* 71(3):677–701.
Benford, Robert D., and David A. Snow. 2000. "Framing Processes and Social Movements: An Overview and Assessment." *Annual Review of Sociology* 26:611–39.
Bennett, W. Lance, and Alexandra Segerberg. 2013. *The Logic of Connective Action: Digital Media and the Personalization of Contentious Politics*. New York: Cambridge University Press.
Bevington, Douglas. 2009. *The Rebirth of Environmentalism: Grassroots Activism from the Spotted Owl to the Polar Bear*. Washington, DC: Island Press.
Blee, Kathleen M. 2012. *Democracy in the Making: How Activist Groups Form*. New York: Oxford University Press.
2013. "How Options Disappear: Causality and Emergence in Grassroots Activist Groups." *American Journal of Sociology* 119(3):655–81.
Bobo, Kim, Jackie Kendall, and Steve Max. 2010. *Organizing for Change: Midwest Academy Manual for Activists*, 4th ed. Santa Ana, CA: Seven Locks Press.
Bosso, Christopher J. 2005. *Environment, Inc.: From Grassroots to Beltway*. Lawrence, KS: University Press of Kansas.
Bradbury, Shelly. 2019. "Trump's Visit Sparks Protests; Arrests Made for Blocking Road." *Pittsburgh Post-Gazette*, October 24, P. A-6.
Braunstein, Ruth, Brad R. Fulton, and Richard L. Wood. 2014. "The Role of Bridging Cultural Practices in Racially and Socioeconomically Diverse Civic Organizations." *American Sociological Review* 79(4):705–25.
Brulle, Robert J. 2000. *Agency, Democracy, and Nature: The U.S. Environmental Movement from a Critical Theory Perspective*. Cambridge, MA: The MIT Press.
2015. "The US National Climate Change Movement." Pp. 146–70 in *Changing Climate Politics*, edited by Yael Wolinsky-Nahmias. Los Angeles: Sage/CQ Press.
Brulle, Robert J., and David N. Pellow. 2006. "Environmental Justice: Human Health and Environmental Inequalities." *Annual Review of Public Health* 27:103–124.
Buechler, Steven M. 1990. *Women's Movements in the United States*. New Brunswick, NJ: Rutgers University Press.
Bullard, Robert D. 1990. *Dumping in Dixie: Race, Class, and Environmental Quality*. Boulder, CO: Westview Press.
1993. *Confronting Environmental Racism: Voices from the Grassroots*. Boston: South End Press.
Bullard, Robert D., and Beverly H. Wright. 1992. "The Quest for Environmental Equity: Mobilizing the African-American Community for Social Change." Pp. 39–49 in *American Environmentalism: The U.S. Environmental Movement, 1970–1990*, edited by Riley E. Dunlap and Angela G. Mertig. New York: Taylor & Francis.

Carroll, William K., and R. S. Ratner. 1996. "Master Framing and Cross-Movement Networking in Contemporary Social Movements." *The Sociological Quarterly* 37(4):601–25.

2001. "Sustaining Oppositional Cultures in 'Post-Socialist' Times: A Comparative Study of Three Social Movement Organizations." *Sociology* 35(3):605–29.

Carson, Rachel. 1962. *Silent Spring*. Boston: Houghton Mifflin.

Carty, Victoria. 2015. *Social Movements and New Technology*. Boulder, CO: Westview Press.

Casner, Nicholas. 2003. "Acid Mine Drainage and Pittsburgh's Water Quality." Pp. 89–109 in *Devastation and Renewal: An Environmental History of Pittsburgh and Its Region*, edited by Joel A. Tarr. Pittsburgh, PA: University of Pittsburgh Press.

Chute, Gwen. 2012. "Meet ESAT: The Allegheny Group's Endangered Species Team." *Allegheny Sierran: Newsletter of the Allegheny Group of the Sierra Club* 30(10):6–7.

Cole, Luke W., and Sheila R. Foster. 2001. *From the Ground Up: Environmental Racism and the Rise of the Environmental Justice Movement*. New York: NYU Press.

Corrigall-Brown, Catherine. 2012. *Patterns of Protest: Trajectories of Participation in Social Movements*. Stanford, CA: Stanford University Press.

Cress, Daniel M., and David A. Snow. 1996. "Mobilization at the Margins: Resources, Benefactors, and the Viability of Homeless Social Movement Organizations." *American Sociological Review* 61(6):1089–109.

Crompton, Janice. 2013. "Activists Oppose Shale Drilling in Pittsburgh Region." *Pittsburgh Post-Gazette*, January 28.

Croteau, David, and Lyndsi Hicks. 2003. "Coalition Forming and the Challenge of a Consonant Frame Pyramid: The Case of a Collaborative Response to Homelessness." *Social Problems* 50(2):251–72.

DeMarco, Patricia M. 2017. *Pathways to Our Sustainable Future: A Global Perspective from Pittsburgh*. Pittsburgh, PA: University of Pittsburgh Press.

Detwiler, John. 2014. "Protect Our Parks." *Allegheny Sierran: Newsletter of the Allegheny Group of the Sierra Club* Fall/Winter:3,5.

Devall, Bill. 1992. "Deep Ecology and Radical Environmentalism." Pp. 51–62 in *American Environmentalism: The U.S. Environmental Movement, 1970–1990*, edited by Riley E. Dunlap and Angela G. Mertig. New York: Taylor & Francis.

Doherty, Brian, and Graeme Hayes. 2019. "Tactics and Strategic Action." Pp. 271–88 in *The Wiley Blackwell Companion to Social Movements*, 2nd ed., edited by Sarah A. Soule, David A. Snow, Hanspeter Kriesi, and Holly J. McCammon. Oxford: John Wiley & Sons.

Dokshin, Fedor A. 2016. "Whose Backyard and What's at Issue? Spatial and Ideological Dynamics of Local Opposition to Fracking in New York State, 2010 to 2013." *American Sociological Review* 81(5):921–48.

Donovan, Brian L. 1995. "Framing and Strategy: Explaining Differential Longevity in the Woman's Christian Temperance Union and the Anti-Saloon League." *Sociological Inquiry* 65(2):143–55.

Dowie, Mark. 1995. *Losing Ground: American Environmentalism at the Close of the Twentieth Century*. Cambridge, MA: The MIT Press.

Downton, James V., Jr, and Paul E. Wehr. 1991. "Peace Movements: The Role of Commitment and Community in Sustaining Member Participation." *Research in Social Movements, Conflicts and Change* 13:113–34.

1997. *The Persistent Activist: How Peace Commitment Develops and Survives*. Boulder, CO: Westview Press.

Dunlap, Riley E., and Angela G. Mertig. 1992. "The Evolution of the U.S. Environmental Movement from 1970 to 1990: An Overview." Pp. 1–10 in *American Environmentalism: The U.S. Environmental Movement, 1970–1990*, edited by Riley E. Dunlap and Angela G. Mertig. New York: Taylor & Francis.

Earl, Jennifer, and Katrina Kimport. 2011. *Digitally Enabled Social Change: Activism in the Internet Age*. Cambridge, MA: MIT Press.

Fawcett, Eliza. 2017. "Residents in Beaver County Worried Shell Pipelines Could Endanger Their Water Supply." *Pittsburgh Post-Gazette*, July 19, Pp. B1, B2.

Ferguson, Cody. 2015. *This Is Our Land: Grassroots Environmentalism in the Late Twentieth Century*. New Brunswick, NJ: Rutgers University Press.

Finney, Carolyn. 2014. *Black Faces, White Spaces: Reimagining the Relationship of African Americans to the Great Outdoors*. Chapel Hill, NC: The University of North Carolina Press.

Fominaya, Cristina Flesher. 2010. "Creating Cohesion from Diversity: The Challenge of Collective Identity Formation in the Global Justice Movement." *Sociological Inquiry* 80(3):377–404.

Freeman, Jo. 1972. "The Tyranny of Structurelessness." Pp. 285–99 in *Radical Feminism*, edited by Anne Koedt, Ellen Levine, and Anita Rapone. New York: Quadrangle Press.

1975. *The Politics of Women's Liberation*. New York: Longman.

1979. "Resource Mobilization and Strategy: A Model for Analyzing Social Movement Organization Actions." Pp. 167–89 in *The Dynamics of Social Movements: Resource Mobilization, Social Control, and Tactics*, edited by Mayer N. Zald and John D. McCarthy. Cambridge, MA: Winthrop.

Freudenberg, Nicholas, and Carol Steinsapir. 1992. "Not in Our Backyards: The Grassroots Environmental Movement." Pp. 27–37 in *American Environmentalism: The U.S. Environmental Movement, 1970–1990*, edited by Riley E. Dunlap and Angela G. Mertig. New York: Taylor & Francis.

Galli, Anya M. 2016. "How Glitter Bombing Lost Its Sparkle: The Emergence and Decline of a Novel Social Movement Tactic." *Mobilization: An International Quarterly* 21(3):259–81.

Gamson, William A. 1990. *The Strategy of Social Protest*, 2nd ed. Belmont, CA: Wadsworth.

Ganz, Marshall. 2000. "Resources and Resourcefulness: Strategic Capacity in the Unionization of California Agriculture, 1959–1966." *American Journal of Sociology* 105(4):1003–62.

2004. "Why David Sometimes Wins: Strategic Capacity in Social Movements." Pp. 177–98 in *Rethinking Social Movements*, edited by Jeff Goodwin and James M. Jasper. Lanham, MD: Rowman & Littlefield.

2009. *Why David Sometimes Wins: Leadership, Organization and Strategy in the California Farm Worker Movement*. New York: Oxford University Press.

Ganz, Marshall, and Elizabeth McKenna. 2019. "Bringing Leadership Back In." Pp. 185–202 in *The Wiley Blackwell Companion to Social Movements, 2nd Edition*, edited by David A. Snow, Sarah A. Soule, Hanspeter Kriesi, and Holly J. McCammon. Hoboken, NJ: John Wiley & Sons Ltd.

Ganz, Marshall, and Ruth Wageman. 2008. "Sierra Club Leadership Development Project: Pilot Project Report and Recommendations." National Sierra Club.

Gerhards, Jurgen, and Dieter Rucht. 1992. "Mesomobilization: Organizing and Framing in Two Protest Campaigns in West Germany." *American Journal of Sociology* 98(3):555–95.

Gottlieb, Robert. 2005. *Forcing the Spring: The Transformation of the American Environmental Movement*. Revised and Updated Edition. Washington, DC: Island Press.

Gould, Deborah B. 2009. *Moving Politics: Emotion and ACT UP's Fight against AIDS*. Chicago: University of Chicago Press.

Gould, Kenneth A., Allan Schnaiberg, and Adam S. Weinberg. 1996. *Local Environmental Struggles: Citizen Activism in the Treadmill of Production*. New York: Cambridge University Press.

Graham, Max. 2018. "Consol Continues Ryerson Longwall Mining." *Pittsburgh Post-Gazette*, August 5.

Griswold, Eliza. 2017. "Undermined: A Local Activist Fights for the Future of Coal Country." *The New Yorker*, July 3. Pp. 48–57.

2018. *Amity and Prosperity: One Family and the Fracturing of America*. New York: Garrar, Straus and Giroux.

Gullion, Jessica Smartt. 2015. *Fracking the Neighborhood: Reluctant Activists and Natural Gas Drilling*. Cambridge, MA: MIT Press.

Han, Hahrie. 2014. *How Organizations Develop Activists: Civil Associations and Leadership in the 21st Century*. New York: Oxford University Press.

Han, Hahrie, Kenneth T. Andrews, Marshall Ganz, Matthew Baggetta, and Chaeyoon Lim. 2011. "The Relationship of Leadership Quality to the Political Presence of Civic Associations." *Perspectives on Politics* 9(1):45–59.

Han, Hahrie, Aaron C. Sparks, and Nate Deshmukh Towery. 2017. "Opening Up the Black Box: Citizen Group Strategies for Engaging Grassroots Activism in the Twenty-First Century." *Interest Groups & Advocacy* 6(1):22–43.

Hays, Samuel P. 2003. "Beyond Celebration: Pittsburgh and Its Region in the Environmental Era—Notes by a Participant Observer." Pp. 193–215 in *Devastation and Renewal: An Environmental History of Pittsburgh and Its Region*, edited by Joel A. Tarr. Pittsburgh, PA: University of Pittsburgh Press.

1979. "*Conservation and the Gospel of Efficiency: The Progressive Conservative Movement, 1890–1920*." New York: Atheneum.

Hirsch, Eric L. 1990. "Sacrifice for the Cause: Group Processes, Recruitment, and Commitment in a Student Social Movement." *American Sociological Review* 55(2):243–54.

Hochschild, Arlie Russell. 2016. *Strangers in Their Own Land: Anger and Mourning on the American Right*. New York: New Press.

Hopey, Don. 2013a. High Court Sides with Towns over Gas Drilling: Act Overriding Local Zoning Laws Is Struck Down." *Pittsburgh Post-Gazette*, December 20, Pp. A-1, A-5.

2013b. "Pennsylvania Environmental Secretary to Step Down." *Pittsburgh Post-Gazette*, March 23, Pp. A1, A3.

2017a. "Gore: Pittsburgh 'Leading the Way' on Climate." *Pittsburgh Post-Gazette*, October 18, Pp. A1, A2.

2017b. "Activists March against Planned Shell Plant in Beaver County: Fracking Protesters Target Conference." *Pittsburgh Post-Gazette*, June 21, Pp. B1, B3.

Hopey, Don, and David Templeton. 2019. "Drilling Critic Testifies before State Criminal Grand Jury." *Pittsburgh Post-Gazette*, February 21, Pp. B1, B3.

Hopkins, Rob. 2008. *The Transition Handbook: From Oil Dependency to Local Resilience*. White River Junction, VT: Chelsea Green Publishing Company.

2011. *The Transition Companion: Making Your Community More Resilient in Uncertain Times*. White River Junction, VT: Chelsea Green Publishing Company.

Jasper, James M. 2004. "A Strategic Approach to Collective Action: Looking for Agency in Social-Movement Choices." *Mobilization* 9(1):1–16.

2006. *Getting Your Way: Strategic Dilemmas in Real Life*. Chicago: University of Chicago Press.

2012. "Choice Points, Emotional Batteries, and Other Ways to Find Strategic Agency at the Microlevel." Pp. 23–42 in *Strategies for Social Change*, edited by Gregory M. Maney, Rachel V. Kutz-Flamenbaum, Deana A. Rohlinger, and Jeff Goodwin. Minneapolis: University of Minnesota Press.

Johnson, Erik W., and Scott Frickel. 2011. "Ecological Threat and the Founding of US National Environmental Movement Organizations, 1962–1998." *Social Problems* 58(3):305–29.

Killian, Lewis M. 1984. "Organization, Rationality and Spontaneity in the Civil Rights Movement." *American Sociological Review* 49:770–83.

Klandermans, Bert. 1992. "The Social Construction of Protest and Multiorganizational Fields." in *Frontiers in Social Movement Theory*, edited by Aldon D. Morris and Carol McClurg Mueller. New Haven, CT: Yale University Press.

Kline, Benjamin. 2011. *First Along the River: A Brief History of the U.S. Environmental Movement*. Lanham, MD: Rowman & Littlefield.

Kretschmer, Kelsy. 2019. *Fighting for NOW: Diversity and Discord in the National Organization for Women*. Minneapolis: University of Minnesota Press.

Kubal, Timothy J. 1998. "The Presentation of Political Self: Cultural Resonance and the Construction of Collective Action Frames." *The Sociological Quarterly* 39(4):539–54.

Ladd, Anthony E., ed. 2018. *Fractured Communities: Risk, Impacts, and Protest against Hydraulic Fracking in U.S. Shale Regions*. New Brunswick, NJ: Rutgers University Press.

Leach, Darcy K. 2016. "When Freedom Is Not an Endless Meeting: A New Look at Consensus-Based Decision Making." *The Sociological Quarterly* 57:36–70.

Legere, Laura. 2017a. "Judges: Coal Project Would Destroy Stream." *Pittsburgh Post-Gazette*, August 16, Pp. A1–A2.

2017b. "Judge Pauses Pipeline Drilling." *Pittsburgh Post-Gazette*, July 26, Pp. E1–E2.

Legere, Laura, and Don Hopey. 2017. "Pa. Supreme Court Rules State Forest Gas Royalty Transfers Unconstitutional." *Pittsburgh Post-Gazette*, June 21, Pp. A1, A8.

Lesle, Timothy. 2006. "States Take Lead on Mercury, Global Warming." Sierra Club, The Planet Newsletter, July/August. https://vault.sierraclub.org/planet/backissues.asp

Levi, Margaret, and Gillian H. Murphy. 2006. "Coalitions of Contention: The Case of the WTO Protests in Seattle." *Political Studies* 54:651–70.

Lichterman, Paul. 1995. "Piecing Together Multicultural Community: Cultural Differences in Community Building among Grass-Roots Environmentalists." *Social Problems* 42(4):513–34.

1996. *The Search for Political Community: American Activists Reinventing Commitment.* Cambridge: Cambridge University Press.
1999. "Talking Identity in the Public Sphere: Broad Visions and Small Spaces in Sexual Identity Politics." *Theory and Society* 28:101–41.
2005. *Elusive Togetherness: Church Groups Trying to Bridge America's Divisions.* Princeton, NJ: Princeton University Press.
Longhurst, James. 2010. *Citizen Environmentalists.* Medford, MA: Tufts University Press.
Lord, Rich. 2010. "Council Hears Sharp Warnings against Marcellus Shale Drilling." *Pittsburgh Post-Gazette*, October 19, Pp. B1, B2.
2013. "Eminent Domain Plays a Key Role in Pipeline Projects." *Pittsburgh Post-Gazette*, April 27, Pp. A1, A6.
Luders, Joseph E. 2010. *The Civil Rights Movement and the Logic of Social Change.* New York: Cambridge University Press.
Maney, Gregory M., Kenneth T. Andrews, Rachel V. Kutz-Flamenbaum, Deana A. Rohlinger, and Jeff Goodwin. 2012. "An Introduction to Strategies for Social Change." Pp. xi–xxxviii in *Strategies for Social Change*, edited by Gregory M. Maney, Rachel V. Kutz-Flamenbaum, Deana A. Rohlinger, and Jeff Goodwin. Minneapolis: University of Minnesota Press.
Marwell, Gerald, and Pamela Oliver. 1984. "Collective Action Theory and Social Movements Research." Pp. 1–27 in *Research in Social Movements, Conflicts and Change*, edited by Louis Kriesberg. Greenwich, CT: JAI Press.
Mayer, Brian. 2009. "Cross-Movement Coalition Formation: Bridging the Labor-Environment Divide." *Sociological Inquiry* 79(2):219–39.
Mayer, Brian, Phil Brown, and Rachel Morello-Frosch. 2010. "Labor-Environmental Coalition Formation: Framing and the Right to Know." *Sociological Forum* 25(4):746–68.
McAdam, Doug. 1983. "Tactical Innovation and the Pace of Insurgency." *American Sociological Review* 48(6):735–54.
2017. "Social Movement Theory and the Prospects for Climate Change Activism in the United States." *Annual Review of Political Science* 20:189–208.
McAdam, Doug, and Hilary S. Boudet. 2012. *Putting Movements in Their Place: Explaining Variation in Community Response to the Siting of Proposed Energy Projects.* New York: Cambridge University Press.
McAdam, Doug, John D. McCarthy, and Mayer N. Zald. 1988. "Social Movements." Pp. 695–737 in *Handbook of Sociology*, edited by Neil J. Smelser. Newbury Park, CA: Sage.
McCammon, Holly J. 2012. *The U.S. Women's Jury Movements and Strategic Adaptation: A More Just Verdict.* New York: Cambridge University Press.
McCammon, Holly J., Lyndi Hewitt, and Sandy Smith. 2004. ""No Weapon Save Argument": Strategic Frame Amplification in the U.S. Woman Suffrage Movements." *The Sociological Quarterly* 45(3):529–56.
McCarthy, John D. 1996. "Constraints and Opportunities in Adopting, Adapting, and Inventing." Pp. 141–51 in *Comparative Perspectives on Social Movements*, edited by Doug McAdam, John D. McCarthy, and Mayer N. Zald. New York: Cambridge University Press.
McCarthy, John D., and Mark Wolfson. 1996. "Resource Mobilization by Local Social Movement Organizations: Agency, Strategy, and Organization in the Movement against Drinking and Driving." *American Sociological Review* 61(6):1070–88.

McCarthy, John D., and Mayer N. Zald. 1977. "Resource Mobilization and Social Movements: A Partial Theory." *American Journal of Sociology* 82(6):1212–41.

McCarthy, John D., David W. Britt, and Mark Wolfson. 1991. "The Institutional Channeling of Social Movements by the State in the United States." *Research in Social Movements, Conflicts and Change* 13:45–76.

McGill, Andrew. 2013. "County Considers Drilling Beneath Parks: Deer Lakes Park Is the First to Be Negotiated in Deal." *Pittsburgh Post-Gazette*, June 13. Pp. B-1, B-2.

McGurty, Eileen. 2007. *Transforming Environmentalism: Warren County, PCBs, and the Origins of Environmental Justice*. New Brunswick, NJ: Rutgers University Press.

McKenna, Elizabeth, and Hahrie Han. 2014. Groundbreakers: How Obama's 2.2 Million Volunteers Transformed Campaigning in America. New York: Oxford University Press.

McKibben, Bill. 2013. "The Fossil Fuel Resistance." Pp. 40–47 in *Rolling Stone*, April 25.

Mershon, Sherie A., and Joel A. Tarr. 2003. "Strategies for Clean Air: The Pittsburgh and Allegheny County Smoke Control Movements, 1940–1960." Pp. 145–73 in *Devastation and Renewal: An Environmental History of Pittsburgh and Its Region*. Pittsburgh, PA: University of Pittsburgh Press.

Meyer, David S., and Deana A. Rohlinger. 2012. "Big Books and Social Movements: A Myth of Ideas and Social Change." *Social Problems* 59(1):136–53.

Meyer, David S., and Suzanne Staggenborg. 2012. "Thinking About Strategy." Pp. 3–22 in *Strategies for Social Change*, edited by Gregory M. Maney, Rachel V. Kutz-Flamenbaum, Deana A. Rohlinger, and Jeff Goodwin. Minneapolis: University of Minnesota Press.

Meyer, David S., and Sidney Tarrow, eds. 2018. *The Resistance: The Dawn of the Anti-Trump Opposition Movement*. New York: Oxford University Press.

Meyer, Rachel, and Howard Kimeldorf. 2015. "Eventful Subjectivity: The Experiential Sources of Solidarity." *Journal of Historical Sociology* 28(4):429–57.

Mitchell, Robert Cameron. 1989. "From Conservation to Environmental Movement: The Development of the Modern Environmental Lobbies." Pp. 81–113 in *Government and Environmental Politics: Essays on Historical Developments Since World War Two*, edited by Michael J. Lacey. Washington, DC: The Wilson Center Press.

Mitchell, Robert Cameron, Angela G. Mertig, and Riley E. Dunlap. 1992. "Twenty Years of Environmental Mobilization: Trends Among National Environmental Organizations." Pp. 11–26 in *American Environmentalism: The U.S. Environmental Movement, 1970–1990*, edited by Riley E. Dunlap and Angela G. Mertig. Washington, DC: Taylor & Francis.

Mohai, Paul, David Pellow, and J. Timmons Roberts. 2009. "Environmental Justice." *Annual Review of Environment and Resources* 34:405–30.

Noakes, John A., and Hank Johnston. 2005. "Frames of Protest: A Road Map to a Perspective." Pp. 1–29 in *Frames of Protest: Social Movements and the Framing Perspective*, edited by Hank Johnston and John A. Noakes. Lanham, MD: Rowman & Littlefield.

Nobel, Justin. 2017. "The Small Town that Fought Fracking." Pp. 38–43, 54 in *Rolling Stone*, June 1.

Norgaard, Kari Marie. 2011. *Living in Denial: Climate Change, Emotions, and Everyday Life*. Cambridge, MA: MIT Press.

Obach, Brian K. 2004. *Labor and the Environmental Movement: The Quest for Common Ground*. Cambridge, MA: MIT Press.

Ogrodnik, Corinne, and Suzanne Staggenborg. 2016. "The Ebb and Flow of Environmentalism." *Sociology Compass* 10(3):218–29.

Pellow, David Naguib. 2007. *Resisting Global Toxics: Transnational Movements for Environmental Justice*: Cambridge, MA: MIT Press.

2014. *Total Liberation: The Power and Promise of Animal Rights and the Radical Earth Movement*. Minneapolis: University of Minnesota Press.

PennEnvironment Research & Policy Center. 2014. *Frack Attack: Pennsylvania's Parks and Forests Threatened by Drilling*. https://pennenvironment.org/reports/pae/frack-attack-pennsylvanias-parks-and-forests-threatened-drilling

Piven, Frances Fox, and Richard A. Cloward. 1977. *Poor People's Movements: Why They Succeed, How They Fail*. New York: Vintage Books.

Polletta, Francesca. 1999. "'Free Spaces' in Collective Action." *Theory and Society* 28:1–38.

2002. *Freedom Is an Endless Meeting: Democracy in American Social Movements*. Chicago: University of Chicago Press.

2012. "Three Mechanisms by Which Culture Shapes Movement Strategy: Repertoires, Institutional Norms, and Metonymy." Pp. 43–57 in *Strategies for Social Change*, edited by Gregory M. Maney, Rachel V. Kutz-Flamenbaum, Deana A. Rohlinger, and Jeff Goodwin. Minneapolis: University of Minnesota Press.

Potter, Chris. 2017. "Pittsburgh Activists Join Dakota Pipeline Protest." *Pittsburgh Post-Gazette*, January 1, Pp. A1, A8.

Rabinowitz, Joni, and John Detwiler. 2014. "Common Sense for Parks; See How Deer Lakes Fracking Goes before Any Expansion." *Pittsburgh Post-Gazette*, December 11, P. B-7.

Reger, Jo, and Suzanne Staggenborg. 2006. "Patterns of Mobilization in Local Movement Organizations: Leadership and Strategy in Four National Organization for Women Chapters." *Sociological Perspectives* 49(3):297–323.

Resnick-Day, Eva. 2017. "Pittsburgh's Ready for 100 off to a Galloping Start." *Allegheny Sierran: Newsletter of the Allegheny Group of the Sierra Club* 38(2):6–8.

Riely, Kaitlynn. 2014. "Allegheny County Begins Health Assessment Meetings Tonight." *Pittsburgh Post-Gazette*, September 22.

Rohlinger, Deana A. 2002. "Framing the Abortion Debate: Organizational Resources, Media Strategies, and Movement-Countermovement Dynamics." *The Sociological Quarterly* 43(4):479–507.

2015. *Abortion Politics, Mass Media, and Social Movements in America*. New York: Cambridge University Press.

Rome, Adam. 2003. "'Give Earth a Chance': The Environmental Movement and the Sixties." *The Journal of American History* 90(2):525–54.

2013. *The Genius of Earth Day*. New York: Hill and Wang.

Rose, Fred. 2000. *Coalitions Across the Class Divide: Lessons from the Labor, Peace, and Environmental Movements*. Ithaca, NY: Cornell University Press.

1997. "Toward a Class-Cultural Theory of Social Movements: Reinterpreting Social Movements." *Sociological Forum* 12:461–94.

Rosenthal, Naomi, and Michael Schwartz. 1989. "Spontaneity and Democracy in Social Movements." *International Social Movement Research* 2:33–59.
Rossi, Federico M. 2017. *The Poor's Struggle for Political Incorporation: The Piquetero Movement in Argentina*. Cambridge and New York: Cambridge University Press.
Roth, Benita. 2010. "'Organizing One's Own' as Good Politics: Second Wave Feminists and the Meaning of Coalition." Pp. 99–118 in *Strategic Alliances: Coalition Building and Social Movements*, edited by Nella Van Dyke and Holly J. McCammon. Minneapolis, University of Minnesota Press.
Roth, Silke. 2003. *Building Movement Bridges: The Coalition of Labor Union Women*. Westport, CT: Praeger.
2008. "Dealing with Diversity: The Coalition of Labor Union Women." Pp. 213–31 in *Identity Work in Social Movements*, edited by Jo Reger, Daniel J. Myers, and Rachel L. Einwohner. Minneapolis: University of Minnesota Press.
Routh, Julian. 2019a. "Trump Likens Shale Gas to 'Gold' in Touting Local Industrial Revival." *Pittsburgh Post-Gazette*, August 14, Pp. A1, A4.
2019b. "Trump Decries Climate Accord in Pittsburgh." *Pittsburgh Post-Gazette*, October 24, Pp. A1, A6.
Sale, Kirkpatrick. 1993. *The Green Revolution: The American Environmental Movement 1962–1992*. New York: Hill and Wang.
Schlosberg, David. 2007. *Defining Environmental Justice: Theories, Movements, and Nature*. Oxford: Oxford University Press.
Seelye, Katharine Q. 2011. "Five Die as Blast Levels Pennsylvania Homes." New York Times, February 11, P. A13.
Shabecoff, Philip. 2003. *A Fierce Green Fire*, rev. ed. Washington, DC: Island Press.
Shalefield Stories. 2012. Homestead, PA: Steel Valley Printers.
Shalefield Stories, vol. 2. 2015. Pittsburgh, PA: Friends of the Harmed.
Shellenberger, Michael, and Ted Nordhaus. 2004. "The Death of Environmentalism: Global Warming Politics in a Post-Environmental World." www.thebreakthrough.org/images/Death_of_Environmentalism.pdf
Smydo, Joe. 2010. "City Oks Ban on Gas Drilling: Council Cites Environmental, Health Concerns." *Pittsburgh Post-Gazette*, November 17.
Snow, David A., and Robert D. Benford. 1992. "Ideology, Frame Resonance and Participant Mobilization." *International Social Movement Research* 1:197–217.
Snow, David A., and Dana M. Moss. 2014. "Protest on the Fly: Toward a Theory of Spontaneity in the Dynamics of Protest and Social Movements." *American Sociological Review* 79(6):1122–43.
Snow, David A., Louis A. Zurcher Jr, and Sheldon Ekland-Olson. 1980. "Social Networks and Social Movements: A Microstructural Approach to Differential Recruitment." *American Sociological Review* 45(5):787–801.
Spears, Ellen G. 2020. *Rethinking the American Environmental Movement Post-1945*. New York: Routledge.
Staggenborg, Suzanne. 1989. "Stability and Innovation in the Women's Movement: A Comparison of Two Movement Organizations." *Social Problems* 36(1):75–92.
1993. "Critical Events and the Mobilization of the Pro-Choice Movement." *Research in Political Sociology* 6:319–45.
1998. "Social Movement Communities and Cycles of Protest: The Emergence and Maintenance of a Local Women's Movement." *Social Problems* 45(2):180–204.
2015. "Event Coalitions in the Pittsburgh G20 Protests." *The Sociological Quarterly* 56:386–411.

2018. "Mobilizing against Fracking: Marcellus Shale Protest in Pittsburgh." Pp. 107–27 in *Fractured Communities*, edited by Anthony E. Ladd. New Brunswick, NJ: Rutgers University Press.

Staggenborg, Suzanne, and Josée Lecomte. 2009. "Social Movement Campaigns: Mobilization and Outcomes in the Montreal Women's Movement Community." *Mobilization* 14(2):405–22.

Staggenborg, Suzanne, and Corinne Ogrodnik. 2015. "New Environmentalism and Transition Pittsburgh." *Environmental Politics* 24(5):723–41.

Strobel, Margaret. 1995. "Organizational Learning in the Chicago Women's Liberation Union." Pp. 145–64 in *Feminist Organizations*, edited by Myra Marx Ferree and Patricia Yancey Martin. Philadelphia: Temple University Press.

Summers-Effler, Erika. 2010. *Laughing Saints and Righteous Heroes: Emotional Rhythms in Social Movement Groups*. Chicago: University of Chicago Press.

Szasz, Andrew. 1994. *Ecopopulism: Toxic Waste and the Movement for Environmental Justice*. Minneapolis: University of Minnesota Press.

Tarr, Joel A., ed. 2003. *Devastation and Renewal: An Environmental History of Pittsburgh and Its Region*. Pittsburgh, PA: University of Pittsburgh Press.

Tarrow, Sidney. 2011. *Power in Movement: Social Movements and Contentious Politics, Revised and Updated*, 3rd ed. New York: Cambridge University Press.

2013. *The Language of Contention: Revolutions in Words 1688–2012*. New York: Cambridge University Press.

Taylor, Dorceta E. 2014. "The State of Diversity in Environmental Organizations." Green 2.0. https://www.diversegreen.org/wp-content/uploads/2015/10/FullReport_Green2.0_FINAL.pdf

2016. *The Rise of the American Conservation Movement: Power, Privilege, and Environmental Protection*. Durham: Duke University Press.

Taylor, Verta. 2000. "Emotions and Identity in Women's Self-Help Movements." Pp. 271–99 in *Self, Identity, and Social Movements*, edited by Sheldon Stryker, Timothy J. Owens and Robert W. White. Minneapolis: University of Minnesota Press.

Taylor, Verta, and Nella Van Dyke. 2004. "'Get Up, Stand Up': Tactical Repertoires of Social Movements." Pp. 262–93 in *The Blackwell Companion to Social Movements*, edited by David A. Snow, Sarah A. Soule, and Hanspeter Kriesi. Oxford: Blackwell.

Taylor, Verta, and Nancy E. Whittier. 1992. "Collective Identity in Social Movement Communities: Lesbian Feminist Mobilization." Pp. 104–29 in *Frontiers in Social Movement Theory*, edited by Aldon D. Morris and Carol McClurg Mueller. New Haven, CT: Yale University Press.

Taylor, Wendi and Phil Coleman. 2016. "Forest Watch Could Help Extend Our Reach." *The Sylvanian: Pennsylvania Chapter Sierra Club*, Spring, p. 2.

Tilly, Charles. 1978. *From Mobilization to Revolution*. Reading, MA: Addison-Wesley Publishing Company.

2008. *Contentious Performances*. New York: Cambridge University Press.

Tilly, Charles, and Lesley J. Wood. 2013. *Social Movements, 1768–2012*, 3rd ed. Boulder, CO: Paradigm Publishers.

Urbina, Ian. 2011a. "Insiders Sound an Alarm Amid a Natural Gas Rush." *The New York Times*, June 26. Pp. A1, A22.

2011b. "Gas Wells Recycle Water, but Toxic Risks Persist." *The New York Times*, March 2. Pp. A1, A14.

Vasi, Ion Bogdan, Edward T. Walker, John S. Johnson, and Hui Fen Tan. 2015. "'No Fracking Way!' Documentary Film, Discursive Opportunity, and Local Opposition against Hydraulic Fracturing in the United States, 2010 to 2013." *American Sociological Review* 80(5):934–59.

Walker, Edward T., and Lina M. Stepick. 2014. "Strength in Diversity? Group Heterogeneity in the Mobilization of Grassroots Organizations." *Sociology Compass* 8(7):959–75.

Walsh, Edward J. 1988. *Democracy in the Shadows: Citizen Mobilization in the Wake of the Accident at Three Mile Island.* Westport, CT: Greenwood Press.

Walsh, Edward J., Rex Warland, and D. Clayton Smith. 1997. *Don't Burn It Here: Grassroots Challenges to Trash Incinerators.* University Park, PA: Pennsylvania State University Press.

Wang, Dan J., and Sarah A. Soule. 2012. "Social Movement Organizational Collaboration: Networks of Learning and the Diffusion of Protest Tactics, 1960–1995." *American Journal of Sociology* 117(6):1674–722.

2016. "Tactical Innovation in Social Movements: The Effects of Peripheral and Multi-Issue Protest." *American Sociological Review* 81(3):517–48.

Ward, Jane. 2008. "Diversity Discourse and Multi-Identity Work in Lesbian and Gay Organizations." Pp. 233–55 in *Identity Work in Social Movements*, edited by Jo Reger, Daniel J. Myers, and Rachel L. Einwohner. Minneapolis: University of Minnesota Press.

Whittier, Nancy. 1995. *Feminist Generations: The Persistence of the Radical Women's Movement.* Philadelphia: Temple University Press.

1997. "Political Generations, Micro-Cohorts, and the Transformation of Social Movements." *American Sociological Review* 62(5):760–78.

Wilber, Tom. 2012. *Under the Surface: Fracking, Fortunes, and the Fate of the Marcellus Shale.* Ithaca, NY: Cornell University Press.

Woliver, Laura R. 1993. *From Outrage to Action: The Politics of Grass-Roots Dissent.* Urbana, IL: University of Illinois Press.

1996. "Mobilizing and Sustaining Grassroots Dissent." *Journal of Social Issues* 52(1):139–51.

Wood, Lesley J. 2012. *Direct Action, Deliberation, and Diffusion.* New York: Cambridge University Press.

Wood, Lesley J., Suzanne Staggenborg, Glenn J. Stalker, and Rachel V. Kutz-Flamenbaum. 2017. "Eventful Events: Local Outcomes of G20 Summit Protests in Pittsburgh and Toronto." *Social Movement Studies* 16(5):595–609.

Woodhouse, Keith M. 2018. *The Ecocentrists: A History of Radical Environmentalism.* New York: Columbia University Press.

Wray, Peter. 2010. "Allegheny Group: A Forty Year History." April 14. Unpublished.

2016. "Bruce Sundquist: Scientist, Outings Leader, Publisher, and Conservationist, 1936–2016." *Allegheny Sierran: Newsletter of the Allegheny Group of the Sierra Club* 36(2):5.

Yonek, Kenneth. 2015. "Environmental Justice Activities with NAACP in the Monongahela River Valley." *Allegheny Sierran: Newsletter of the Allegheny Group of the Sierra Club* 33(1):1,4.

Zald, Mayer N., and John D. McCarthy. 1987. *Social Movements in an Organizational Society: Collected Essays.* New Brunswick, NJ: Transaction Books.

Index

Books in the Series (*continued from p.iii*)

Lars-Erik Cederman, Kristian Skrede Gleditsch, and Halvard Buhaug, *Inequality, Grievances, and Civil War*
Christian Davenport, *How Social Movements Die: Repression and Demobilization of the Republic of New Africa*
Christian Davenport, *Media Bias, Perspective, and State Repression*
Gerald F. Davis, Doug McAdam, W. Richard Scott, and Mayer N. Zald, *Social Movements and Organization Theory*
Donatella della Porta, *Clandestine Political Violence*
Donatella della Porta, *Where Did the Revolution Go? Contentious Politics and the Quality of Democracy*
Mario Diani, *The Cement of Civil Society: Studying Networks in Localities*
Nicole Doerr, *Political Translation: How Social Movement Democracies Survive*
Barry Eidlin, *Labor and the Class Idea in the United States in Canada*
Todd A. Eisenstadt, *Politics, Identity, and Mexico's Indigenous Rights Movements*
Grzegorz Ekiert, Elizabeth J. Perry, and Yan Xiaojun editors, *Ruling by Other Means: State-Mobilized Movements*
Olivier Fillieule and Erik Neveu, editors, *Activists Forever? Long-Term Impacts of Political Activism*
Diana Fu, *Mobilizing Without the Masses: Control and Contention in China*
Daniel Q. Gillion, *The Political Power of Protest: Minority Activism and Shifts in Public Policy*
Marco Giugni and Maria Grasso *Street Citizens: Protest Politics and Social Movement Activism in the Age of Globalization*
Jack A. Goldstone, editor, *States, Parties, and Social Movements*
Jennifer Hadden, *Networks in Contention: The Divisive Politics of Climate Change*
Michael T. Heaney and Fabio Rojas, *Party in the Street: The Antiwar Movement and the Democratic Party after 9/11*
Tamara Kay, *NAFTA and the Politics of Labor Transnationalism*
Neil Ketchley, *Egypt in a Time of Revolution: Contentious Politics and the Arab Spring*
Joseph Luders, *The Civil Rights Movement and the Logic of Social Change*
Doug McAdam and Hilary Boudet, *Putting Social Movements in Their Place: Explaining Opposition to Energy Projects in the United States, 2000–2005*
Doug McAdam, Sidney Tarrow, and Charles Tilly, *Dynamics of Contention*
Holly J. McCammon, *The U.S. Women's Jury Movements and Strategic Adaptation: A More Just Verdict*
Sharon Nepstad, *Religion and War Resistance and the Plowshares Movement*
Olena Nikolayenko, *Youth Movements and Elections in Eastern Europe*
Kevin J. O'Brien and Lianjiang Li, *Rightful Resistance in Rural China*
Silvia Pedraza, *Political Disaffection in Cuba's Revolution and Exodus*
Héctor Perla Jr., *Sandinista Nicaragua's Resistance to US Coercion*
Federico M. Rossi, *The Poor's Struggle for Political Incorporation: The Piquetero Movement in Argentina*
Chandra Russo, *Solidarity in Practice: Moral Protest and the US Security State*
Eduardo Silva, *Challenging Neoliberalism in Latin America*
Erica S. Simmons, *Meaningful Resistance: Market Reforms and the Roots of Social Protest in Latin America*

Sarah Soule, *Contention and Corporate Social Responsibility*
Sherrill Stroschein, *Ethnic Struggle, Coexistence, and Democratization in Eastern Europe*
Yang Su, *Collective Killings in Rural China during the Cultural Revolution*
Sidney Tarrow, *The Language of Contention: Revolutions in Words, 1688–2012*
Sidney Tarrow, *The New Transnational Activism*
Wayne P. Te Brake, *Religious War and Religious Peace in Early Modern Europe*
Ralph A. Thaxton Jr., *Catastrophe and Contention in Rural China: Mao's Great Leap Forward Famine and the Origins of Righteous Resistance in Da Fo Village*
Ralph A. Thaxton Jr., *Force and Contention in Contemporary China: Memory and Resistance in the Long Shadow of the Catastrophic Past*
Charles Tilly, *Contention and Democracy in Europe, 1650–2000*
Charles Tilly, *Contentious Performances*
Charles Tilly, *The Politics of Collective Violence*
Marisa von Bülow, *Building Transnational Networks: Civil Society and the Politics of Trade in the Americas*
Lesley J. Wood, *Direct Action, Deliberation, and Diffusion: Collective Action after the WTO Protests in Seattle*
Stuart A. Wright, *Patriots, Politics, and the Oklahoma City Bombing*
Deborah Yashar, *Contesting Citizenship in Latin America: The Rise of Indigenous Movements and the Postliberal Challenge*
Andrew Yeo, *Activists, Alliances, and Anti-U.S. Base Protests*

CPSIA information can be obtained
at www.ICGtesting.com
Printed in the USA
LVHW030234130821
695223LV00018B/1668

9 781108 745864